Advancing Real Property Valuation

Digital technologies are reshaping the way property and asset values are assessed, regulated and trusted. From automated valuations to drone-assisted site inspections, these advances are influencing professional practice and testing the legal, regulatory and institutional frameworks – including those established by IVSC, API and RICS – that underpin it.

This book examines how AI, drones and emerging digital tools are being integrated into valuation processes across varied contexts. Its scope extends beyond traditional property valuation to urban markets, public assets, infrastructure and ESG assessments. Drawing on insights from leading academics and industry experts in Australia and internationally, it's organised into four sections: digital foundations and frameworks, real-world applications and innovations, advancing education and training and legal, ethical and future perspectives. Across these sections, comparative analysis explores how different jurisdictions and sectors are adapting, with particular attention to regulatory compliance, professional standards and the boundaries of automation.

Rather than framing technology as an inevitable disruption or existential threat, this book encourages careful, critical engagement with how digital systems shape professional judgement, data transparency and institutional trust. It offers theoretical grounding and curricular models that prepare the next generation of valuation professionals to operate in technology-rich environments and provides detailed case studies and practical strategies to uphold standards, manage risk and ensure accountability when using advanced digital tools. It is an essential resource for anyone concerned with the future of valuation, as well as those seeking to upskill and build confidence in using advanced technologies that are transforming property valuation methods and practices.

Johari Hussein Nassor Amar is a Chartered Commercial Property Surveyor and Assistant Professor at Bond University in Australia, where she teaches valuation and various property-related subjects to both undergraduate and postgraduate property degree students. With more than 15 years of experience across academia, research and industry, she applies her commercial property expertise within a human-centric approach to explore the convergence of people, place and policy. This includes specialised work in aged and disability healthcare co-design. Her research and supervision encompass property valuation, affordable housing, built heritage and sustainable property futures, with a focus on integrating emerging technologies such as AI and LiDAR. Her work appears internationally in peer-reviewed academic journals, professional publications and mainstream media. Dr. Amar holds a PhD in Sustainable Development from Bond University, an MSc in Real Estate Management from KTH Royal Institute of Technology in Sweden and a BSc (Hons) in Land Management and Valuation from the University of Dar es Salaam, Tanzania.

Lynne Armitage was an Associate Professor in Urban Development at Bond University. Academically, Dr. Armitage lectured, led and established undergraduate and graduate programmes at UWS, University of Melbourne, QUT, Unitech Lae and Bond University. She has published over 50 refereed journal articles and consultancy reports, written and published four valuation and one land economics textbooks, authored two chapters in a valuation textbook for the American Institute of Appraisers and co-authored three book chapters with Dr Amar on environmental offsets, heritage assets and people/environment fit (P-E Fit) for commercial office design. She held board positions with the Valuers' Queensland Registration Board and the Surveyors' Board of Queensland and chaired the RICS' Oceania Education and Professional Standards Board. In addition, she was an external examiner of RICS- and API-accredited bachelor's and master's degrees in property in Australia, Malaysia and Papua New Guinea.

Advancing Real Property Valuation

AI, Drones and Emerging Digital Technologies in Professional Practice and Education

Edited by Johari Hussein Nassor Amar and Lynne Armitage

Routledge
Taylor & Francis Group
LONDON AND NEW YORK

Designed cover image: iStock by Getty Images

First published 2026
by Routledge
4 Park Square, Milton Park, Abingdon, Oxon OX14 4RN

and by Routledge
605 Third Avenue, New York, NY 10158

Routledge is an imprint of the Taylor & Francis Group, an Informa business

British Library Cataloguing-in-Publication Data
A catalogue record for this book is available from the British Library

ISBN: 978-1-032-87211-7 (hbk)
ISBN: 978-1-032-88707-4 (pbk)
ISBN: 978-1-003-53925-4 (ebk)

DOI: 10.1201/9781003539254

Typeset in Times New Roman
by SPi Technologies India Pvt Ltd (Straive)

In Memoriam
Associate Professor Dr Lynne Armitage
1950–2025

This volume is dedicated to Dr Lynne Armitage, who was an Associate Professor of Urban Development at Bond University, served as co-editor and helped shape the direction of this collection. Her contributions to urban development and property valuation provided the foundation for this work. Following her passing in the first quarter of 2025, colleagues and contributors completed the volume in her honour. While her absence is profoundly felt, her ideas and influence remain present throughout these pages.

Lynne's professional life extended across four decades and several regions, including the United Kingdom, Australia, Southeast Asia, southern Africa and the Pacific. She began her career as a Chartered Surveyor in the UK and later worked across government and private sectors. Her consulting work with the World Bank, AusAID and Ministry of Lands in Thailand offered her a broad view of property systems in varied cultural contexts. This informed her approach to international valuation and land policy.

Her academic work was equally broad and thoughtful. She held teaching and research roles at five universities, including key positions at Hawkesbury Agricultural College (HAC, now University of Western Sydney (UWS)), Papua New Guinea University of Technology (UniTech), Queensland University of Technology (QUT), University of Melbourne, Bond University. She also contributed to developing property education in Papua New Guinea, Thailand, and several island nations. Her published work included journal articles, textbooks, book chapters and consultation reports used widely by students and by professionals. Lynne was a Fellow of the Royal Institution of Chartered Surveyors (FRICS), a Fellow and Certified Practising Valuer of the Australian Property Institute (FAPI CPV), a Registered Valuer with the Queensland Valuers Registration Board and a Chartered Member of the Society for the Environment (CEnv). She served on several professional boards and chaired the RICS Oceania Education and Professional Standards Board.

Those who worked with Lynne will remember her not only for her knowledge, but also for her patience, honesty and commitment to high standards. She took time to support others and was generous with her insights. Her interest in the connections between people, place, property and technology shaped how she approached both teaching and research. Her views often reminded us to look beyond rules and numbers and to consider the broader context.

This volume reflects themes Lynne cared about deeply. Her professional ethics and standards opened space for new ways of thinking in valuation and continues through those she taught, challenged and encouraged. We remain grateful for her friendship, her leadership and her belief in the value of good honest work. All contributors to this volume have dedicated their chapters to Lynne in her honour.

Vale Lynne – a mother, friend, mentor, colleague

Contents

Foreword *x*
Preface and acknowledgements *xiii*
List of contributors *xvi*

1 Valuation at a crossroads: From resistance to integration of human-digital systems in the profession 1
JOHARI HUSSEIN NASSOR AMAR

SECTION 1
Digital foundations and frameworks in valuation practice **15**

2 Emerging AI technologies in the valuation of tangible assets 17
ALLEN J CRAWFORD AND KEVIN BROGAN

3 Artificial intelligence and machine learning in property and real asset valuation: Policies, practices and emerging trends 33
ISARA KHANJANASTHITI, SAMUEL GARDNER AND ARMIN TAKLIF

4 Coevolution of artificial intelligence technology and the valuation profession 57
HAO WU AND RITA WONG

5 Challenges and opportunities of data collection and computational analysis using AI in real estate valuation 72
XIN JANET GE AND PETER RAPTIS

SECTION 2
Applied digital innovations in valuation practice 105

6 Public sector valuations: Current and emerging technologies, and industry readiness 107
ELISE WALLACE, ADAM WALLACE, DAVID LOGUE AND MATTHEW MOORHEAD

7 Airport runway inspections: Transitioning from conventional methods to drone-based solutions 127
ISARA KHANJANASTHITI, ARMIN TAKLIF AND SAMUEL GARDNER

8 Generative AI prompt engineering basics for real estate appraisers 150
KEITH WOLF AND JOHARI HUSSEIN NASSOR AMAR

SECTION 3
Innovating valuation education and training for the digital era 183

9 The new role for academics in advancing artificial intelligence in real estate education 185
STEVEN BOYD

10 The future of valuation education: Incorporating augmented reality tools in training and skill development for valuers 218
MAXIMILIAN NEUGER AND LYNNE ARMITAGE

SECTION 4
Reflections on legal, ethical and future directions in digital valuation 235

11 A(I) legal perspective: Process, procedure, risk and GenAI use 237
LUCY CRADDUCK

12 Reflection on AI, big data and analytics in real estate valuation 247
SIMON BENJAMIN FONTEYN

13 Quantum valuation: Navigating the disruption, seizing the opportunities in the new era 258
GREG SKULMOSKI, ALEX LANG, ASHKAN MEMARI AND CEN YING LEE

14 Valuation beyond the divide: Rethinking professional practice and standards in an era of digital transformation 280
JOHARI HUSSEIN NASSOR AMAR

Index *292*

Foreword

This is a book about hope and opportunity for the valuation profession.

It challenges the premise that the role of the valuer and the role of emerging technology are an *either/or* choice, advocating for and showing how an *as well as* selection may be embraced, with the book focusing on developing complementarity between human expertise and computational capabilities, showing that the future of valuation lies in a balanced partnership between human valuers and intelligent machines.

My own career encompasses at least three rounds of new technology which impacted the valuation profession – the introduction of calculators in the late 1970s and early 1980s, the adoption of spreadsheets and cash flows in the late 1980s and early 1990s and the implementation of software valuation packages and automated valuation models in the late 1990s and earlier in this millennium.

The common trend that emerged from each round of technological innovation was initial, protracted resistance to change by valuers, followed by late-in-the-day, begrudging acceptance driven by client demands.

It does not have to be this way again.

This book provides a clear and risk-averse road map to adoption by the valuation profession of such new technologies as artificial intelligence, drones and light detection and ranging.

Identifying that valuation is at a crossroads, the book prompts a conversation about the role of prevalent technologies through four lenses:

- theoretical foundations;
- practical applications;
- educational implications; and
- future directions.

Artificial intelligence, unlike calculators and spreadsheets, may not require the valuer to exercise judgement. While not yet perfect, artificial intelligence is improving exponentially as data quality and availability increase, such that, at some point in the future, the computer can potentially be trained to exercise judgement.

When that point in the future may arise is unclear, but it provides the valuation profession with an opportunity to reposition itself and its services to deliver that which artificial intelligence cannot. The window of opportunity, therefore, arises for those valuers who embrace artificial intelligence to understand not only what it can do but also what it cannot do and redesign their valuation product accordingly.

While, on the one hand, this may be seen as challenging the fundamental notion of what a valuer is, on the other hand, it allows the valuer to focus on oversight of key judgements, being the highest-order role for the valuer, and positions the valuer for the greatest value add in the valuation process. As the book succinctly observes, this may elevate the role of the valuer as the interpreter and validator of technologically derived insights. Further, if implemented thoughtfully, digital augmentation can indeed transform valuation practice for the better.

Further, several chapters of the book predict that valuers will increasingly focus on heterogeneous, high-risk or judgement-intensive assignments and on providing the ethical, contextual and interpersonal elements of service that machines cannot, summarising that the valuation profession is not being replaced but is being redefined.

As the book explores, the adoption of new technologies raises ethical and regulatory issues that may be considered as two sides of the same coin. Ethical codes promulgated by professional bodies are inextricably linked to the regulations prescribed by such professional bodies and by the government. As professional bodies are composed of valuers as members, the window of opportunity also extends to professional bodies to embrace the direction of the profession and adapt ethical codes and regulations accordingly, which the government may then follow.

A further challenge, considered in the book, is the role of education and continuing professional development in preparing new entrants to the profession and existing members of the profession for the impact of new technologies. While automated valuation modelling has been taught in university courses for 40 years, the inclusion of other building blocks to prepare students for a technology-focused workplace should be both welcomed and encouraged by academics and university management.

However, the application of continuing professional development to reskill existing members of the profession may be somewhat more challenging. As the book notes, such reskilling will be required to address mid-career transitions and to help experienced valuers adapt their established expertise to incorporate new technological capabilities.

The challenge may be that, for reskilling to be effective, it requires valuers to acknowledge the need to be reskilled and to be willing to be reskilled. While the development of reskilling courses is eminently feasible, the requisite cultural change in the valuation profession may be far harder to achieve.

The book helpfully concludes by identifying some broad areas where further research and proactive efforts are needed to ensure technology is integrated in a way that strengthens, rather than undermines, the valuation profession, including:

- ethical and equitable integration of artificial intelligence;
- preserving professional expertise and accountability; and
- embracing emerging technologies and data complexities.

Johari and Lynne are to be commended for bringing together a range of specialist authors to produce a book that catalyses conversations for the benefit of the valuation profession.

Finally, I would like to acknowledge Lynne Armitage's four-decade contribution to property academia. I had the pleasure of sharing hospitality and attending conferences with Lynne around the world, as well as being a member of professional body accreditation panels for courses taught by Lynne. Reviewing assessment scripts by students, Lynne's personality and humour shone through as she memorably struck out the word *landlord* in one student's assessment, jotting *sexist and feudal!* in the margin, while for another, she wrote, *show me your references and I'll show you your marks!* Vale Lynne.

David Parker
Sydney
August 2025

Preface and acknowledgements

This volume emerged from our lived experience as educators, researchers and collaborators within the property profession. In our classrooms, we observed the ongoing tension between foundational valuation principles and the rapid integration of emerging technologies. During the COVID-19 pandemic, when teaching shifted online, we encountered challenges in enabling students to conduct site inspections remotely and demonstrate competence in using property databases for desktop valuation tasks.

At Bond University, we responded by incorporating drones into property inspection subjects and began exploring the potential of generative AI across several areas of the property programme. However, we deliberately excluded its use in valuation-specific subjects. This decision was partly influenced by our early training session with ChatGPT, in which Associate Professor Lynne Armitage prompted Generative AI to value her own property. The response was along these lines: *I am sorry, but I can't provide a formal property valuation. You should consult a registered valuer for an accurate and compliant assessment.* This served as a timely reminder of the professional and regulatory boundaries that remain central to valuation practice.

Our interdisciplinary research, which explores the convergence of people, property, place and technology in the built environment, revealed a critical gap. Technologies such as AI, LiDAR and drone-based imaging were advancing independently, yet professionals lacked a comprehensive framework to understand their combined application in valuation practice. Students and practitioners alike were seeking guidance, but existing resources remained fragmented across different disciplines and use cases. Through classroom experiments with both research-based and conversational AI tools, discussions with industry professionals and ongoing supervision of research projects, we recognised a growing need for more than technical skills. What was required was a deeper understanding of how human expertise and technological capability could work together in meaningful and responsible ways.

This volume represents our attempt to bridge that educational and professional divide. Drawing on our combined experiences, we have assembled a collection that offers both conceptual insight and practical guidance. It is informed by real-world experimentation and guided by a commitment to maintaining the rigour and ethical standards that underpin credible valuation practice.

The editorial process began with an open call for proposals in June 2024, attracting 28 submissions from academics and practitioners across Australia, Asia, Europe and North America. After a rigorous three-stage peer review involving academics, practising valuers and technology experts with backgrounds in either valuation or other property-related disciplines, 20 proposals were accepted. Following author withdrawals, 16 full chapters were submitted and reviewed blind by two to three experts per chapter. Twelve chapters were ultimately selected for inclusion, ensuring the volume retained scholarly rigour while remaining relevant to those navigating valuation practice in real time.

We extend our sincere thanks to all reviewers whose insights strengthened the final contributions. We are particularly grateful to all 14 anonymous reviewers whose feedback shaped the direction and clarity of the work. Also listed below are reviewers in alphabetical order, whom we have permission to publicly acknowledge and whose input was equally vital to the success of the project.

Associate *Professor Dr Abdurrahman Yagmur Toprakli (Gazi University, Turkey); Associate Professor Adrian Lee (Property and Real Estate Discipline Leader, Deakin University, Australia); Allen J Crawford FRICS FREIQ FAPI (Managing Director, Corporate & Commercial Property Advisors, QLD, Australia); Professor Claudio Cacciamani (University of Parma, Italy); Dr David Parker FRICS, FAPI, FAICD, SFFin (Visiting Professor, University of Reading, UK); Drew Castley (Partner, Hall & Wilcox, QLD, Australia); Dwi Setiadi Firmansyah (Section Head, UAV Development and Operation, Indonesia); Dr Isara Khanjanasthiti (Lecturer in Urban and Regional Planning, University of New England, Australia); Professor John Sheehan (Faculty of Society and Design, Bond University, Australia); John Pedersen (Director, JP consulting (London) Limited, UK); Dr Julian (Jules) Congalton CAMA (Principal and Director, Asset Dynamics Limited, New Zealand); Kevin Brogan FAPI FRICS (National Director, Group Risk and Compliance, Herron Todd White (HTW), QLD Australia); Mark Wist FRICS FAPDI (Principal, Property Resolutions, Brisbane QLD, Australia); Dr Olga Kaganova (Consultant, World Bank, USA – Chapter 6); Richard Lysnar AAPI CPV (Director, JLL Value and Risk Advisory QLD, Australia); Professor Roberto Moro-Visconti (Facoltà di Economia, Università Cattolica, Italy); Associate Professor Shumank Deep (IMS Business School, Ghaziabad, India); Simon Fonteyn (Managing Director, IPS Consultants Pty Limited, NSW); Dr Steven Boyd FAPI GAICD (Academic Lead Property Studies, CQUniversity, Australia); Sugandhi Jayaraman (Senior Airport Planner & Lecturer, University of Westminster, UK); Dr Troy Nielson (Adjunct Lecturer, CAHE Griffith University, QLD, Australia); Ty Noble (JLL Head of Public Sector Value and Risk Advisory, South Australia); Associate Professor Xin*

Janet Ge MRICS AAPI (School of Built Environment, University of Technology Sydney, Australia); Dr Zhuoran (Thomas) Zhang (Research Associate, Curtin University, Australia).

This volume would not have been possible without the institutional support of Bond University Faculty of Society and Design, which provided time, resources and encouragement throughout the editorial process. The professional and academic organisations for distributing our call for book chapters to their members, including the Royal Institution of Chartered Surveyors (RICS), the Australian Property Institute (API), the American Real Estate Society (ARES), the European Real Estate Society (ERES), the Pacific Rim Real Estate Society (PRRES), the Chartered Institute of Building (CIOB), buildingSMART and New Planner, to mention a few. A thank you to Dr Rosemarie Rusch, who was part of the editorial team but stepped down in February 2025 following her retirement. Dr Belinda Glynn for undertaking the initial professional proofreading and editing of the chapter. And, most importantly, a special thanks to our families, friends and colleagues for their support, encouragement and patience throughout this process.

We trust that this volume serves as both a practical resource for current practitioners and a foundation for future research. As property valuation continues to evolve in a digital context, the ability of professionals to integrate new technologies while preserving professional judgement will be central to its future. The challenge lies not in choosing between technology and human expertise, but in understanding how the two can be integrated intelligently and ethically to support robust and trustworthy valuation practice.

Dr Johari Hussein Nassor Amar
June 2025

The authors would like to acknowledge that GPT 4.0 API, Co-Pilot and Apple Intelligence were utilised by the authors to simplify technical writing for a general audience and as a proofreading and editing assistant to enhance the clarity of the chapter's writing. Chapter 8 used Julius AI to demonstrate basic prompt engineering methods using a data set curated by the author, generated from Realtor Multiple Listing Services data. Prompts were used requesting the AI to use data science methods to evaluate data imported into the AI system. The objective of the chapter was to demonstrate how a real estate appraiser could use Gen AI as a tool, but also identify errors Gen AI is prone to make and the importance of having a human in the loop to review, validate, and confirm the results from a Gen AI tool. The version of Julius used for his analysis used GPT 4.1 and Claude Opus, which were default models. Chapter 10 utilised Claude Sonnet 4, Anthropic, for structural guidance, proofreading, and final editing suggestions. Chapter 12 utilised GPT 4.0 for quality assurance, checking grammar and expression, and for market research on the Value Australia project. Grammarly Pro was also used on several chapters for proofreading and editing purposes.

Contributors

Johari Hussein Nassor Amar

Dr. Johari Hussein Nassor Amar, MRICS AAPI QPCR AFHEA (Indigenous Knowledges), is a Human-centric Built Environment Specialist and Lecturer at Bond University, where she leads the Sustainable Built Environment Research Group (SBERG). With over 15 years of academic and industry experience, her work focuses on interdisciplinary approaches informed by Person–Environment Fit (PE-Fit) theory to explore the convergence of people, place, property and policy. Her research spans affordable housing models, aged and disability healthcare co-design, and sustainable property futures – including built heritage conservation, commercial real estate, land tenure systems and property decarbonisation. Dr Amar integrates digital innovations such as Gen-AI and LiDAR into her teaching, research and HDR supervision. Her work is published internationally in peer-reviewed journals, books and media, contributing to socially inclusive, commercially viable and forward-looking built environment research.

Lynne Armitage

In memoriam: Associate Professor Dr. Lynne Armitage (FRICS CVS, API CPV) held an academic position at Bond University. Her academic work spanned various facets of valuation, heritage and sustainability concerns in property practice. She led national research projects on built environment education and remained active in advancing professional curriculum development until her last days of service. Her recent interests included the role of education and the application of immersive technologies in the training of valuers. Her work in valuation education, land management and built heritage leaves a lasting legacy to the property profession in Australia and beyond.

Steven Boyd

Dr. Steven Boyd is the Academic Lead of Undergraduate Business Studies within the School of Business and Law at CQUniversity, Queensland, Australia.

Since 2001, he has been active in real estate fund management, valuation, consultancy and education. During the past 20 years, Steven was part of the strategic operations team for a Brisbane-based fund management company before moving to regional Queensland to manage a consultancy and valuation practice and then formally joining academia. Steven's interest in education and technological advances in real estate valuation has been at the core of his practice and scholarship research. He continues to develop education modules for the Australian Property Institute, amongst others, and has a series of artefacts, or interventions, to his name, including a suite of serious games and a mobile application, to enhance the learning and analysis of real estate.

Kevin Brogan

Kevin Brogan is the National Director and leads the Valuation Quality, Risk and Compliance function of Herron Todd White, one of the largest independent property valuation and advisory groups in Australia. He has over 35 years of property valuation experience gained in various roles in the United Kingdom and Australia. He is a former Valuer with the Valuation Office Agency in Bath, City Valuer for Adelaide City Council, Commercial Property Valuer with Colliers International and spent 13 years at Corelogic as Principal, Valuation Policy and Compliance/Residential and Property Market Commentator.

Kevin has served as a member of the Australian Property Institute South Australian Committee and contributes to the drafting of API standards and publications. Kevin has served a term on the RICS Governing Council and the RICS Nominations Committee and is currently a member of the RICS Asia Pacific World Regional Board and Chair of the RICS Australasian Advisory Board.He also participates in the RICS Australia Valuation Advisory Forum and is a media commentator for the RICS Global Property/Construction Monitor report in Australia. Kevin is a passionate advocate for professional valuation standards in Australia and overseas and has contributed articles to the Valuers Bulletin on the subject.

Lucy Cradduck

Dr. Lucy Cradduck is an Associate Professor in the QUT Law School, holds a current Practising Certificate as a Solicitor, and is Chair of the Faculty of Business & Law Academic Misconduct Committee (and previously for the Faculty of Law), which determines allegations of academic misconduct by non-research degree students, including allegations regarding the unauthorised use of AI tools. Her research includes property law more broadly, individuals' engagements with technology, and liability issues for valuation professionals.

Allen J Crawford FRICS FREQ FAPI CPV Specialist Retail Valuer (QLD)

Allen Crawford's career in property commenced in February 1973, undertaking property accounting for the AMP Society. He later moved into property management at Centrepoint in Sydney and then into valuation and project management with AMP Society. He has been a Registered Valuer, originally NSW, since 28 June 1981. Allen is currently the Managing Director of Corporate & Commercial Property Advisors.

He was elevated to the status of Fellow of the Australian Property Institute ("API") in 1994. He was elected a Fellow of the Royal Institution of Chartered Surveyors (RICS) in March 2005. Allen currently serves on the Australian Standards Steering Committee of the API, is deputy chair of the Professional Conduct Committee of the API, is a Member of the Industry Advisory Committee for Bond University, is an external examiner on behalf of RICS for Bond University and guest lectures at Bond University.

Simon Benjamin Fonteyn

Simon Benjamin Fonteyn is an experienced property professional with over 30 years of expertise in retail, medical and childcare property valuations, leasing and market rent determinations. He has been engaged as an expert witness in multiple legal proceedings, providing independent assessments in retail leasing disputes. Simon has an extensive track record in rental determinations, valuation methodologies and commercial lease analytics.

He is the Founder and Managing Director of Intelligent Property Solutions Pty Ltd and has built Leaseinfo®, Australia's largest retail leasing database, providing key market insights for landlords, tenants and valuers. He is a Specialist Retail Valuer, a trusted adviser to government and private clients, and a casual lecturer in Property Economics at the University of Technology Sydney (UTS).

Samuel Gardner

Samuel Gardner is a consultant and registered valuer at Marsh McLennan, with expertise in financial reporting, valuations and risk advisory for specialised property and infrastructure portfolios. This expertise spans a wide array of public and private sectors throughout the Southern Hemisphere. With expertise in procurement, feasibility analysis and infrastructure development, Samuel has led consulting engagements and executed projects for federal government agencies across the South Pacific. His extensive experience in stakeholder management has been instrumental in driving major public-private partnerships for large-scale infrastructure initiatives. Samuel is committed to delivering precise financial valuations and actionable insights across intricate asset portfolios.

Xin Janet Ge

Dr. Xin Janet Ge is an Associate Professor in Property Economics and Development at the University of Technology Sydney. Her research focuses on housing affordability, the housing market, property development finance and investment, and the application of property technologies. Janet has a strong track record in collaborative research for organisations and the property industry and has successfully led several research projects to completion. Janet teaches property investment, finance and valuation and has supervised more than 30 doctoral, master's and honours students' theses. Janet is a member of RICS and API and on the Board of Directors of AsRES.

Isara Khanjanasthiti

Dr. Isara Khanjanasthiti is a Lecturer in Urban and Regional Planning at the University of New England, Australia. His research spans planning around airports, planning systems and governance, housing affordability, smart cities and innovative approaches to university teaching and learning. Dr Khanjanasthiti has contributed to research on cross-border governance, placemaking strategies, active living initiatives, sustainable campus development and knowledge-based urban development. Dr Khanjanasthiti earned his bachelor's, master's (awarded with High Distinction) and doctoral degrees in urban planning from Bond University, where he also spent a decade lecturing, tutoring and conducting interdisciplinary research. He has received multiple academic awards, including the *Minister's Town Planning Prize (Runner-up)* for his master's thesis and the *HASSELL Prize for Best Graduating Undergraduate Student in Urban Planning and Design.*

Alex Lang

Dr. Alex Lang brings 20 years of global industry experience, and his research focuses on integrating external information into companies by designing adequate processes and organisational structures to transform the companies into open organisations. He is interested in the movement away from line-to-project-centred organisations to plan, implement and optimise innovations like quantum technologies. Finally, balancing risks and opportunities in a dynamic environment and employee behaviour is also a new research area.

Cen Ying Lee

Dr. Cen-Ying Lee, PhD, B. QS, MSc. CCM., FHEA, is an academic and researcher at Bond University in Australia. Her primary research focuses on sustainability in the built environment, with expertise in the adoption of emerging technologies. One of her notable publications, 'A Mixed Review of the Adoption of Building Information Modelling (BIM) for Sustainability', was ranked among the top 1% of papers in the engineering field by Essential Science Indicators in 2020. Dr Lee teaches a broad range of construction management courses at both undergraduate and postgraduate levels, integrating research-led teaching with a strong emphasis on technological innovation in the built environment.

Bond University

David Logue

Assistant Professor David Logue AAPI (CPV) and Registered Valuer Queensland is a built environment professional and lecturer in Bond University's Property programmes. After more than 20 years of industry experience as a registered property valuer and urban planner, he commenced a career in academia in 2017. His research covers land use, regional development, agglomeration economies and property markets. His current PhD research explores the highest and best use within urban renewal contexts. David brings his industry experience, alignment with industry practitioners and nagging curiosity to the lecture theatre and to his research topics to provide authentic and practical learning experiences and research outcomes.

Ashkan Memari

Dr. Ashkan Memari, PhD, MEng, BEng, FHEA, is an academic and researcher at Central Queensland University in Australia, with international experience across the Middle East, South Asia and Australia. Ashkan's primary research area revolves around sustainability, and his expertise is underscored by the recognition of one of his articles in the field of sustainable supplier selection, which ranks among the top ten most cited works globally in this field.

Dr. Memari teaches a wide range of Project Management courses catering to both undergraduate and postgraduate students. Ashkan focuses on lean implementation tools and techniques, and he has recently contributed to the academic landscape with his article entitled *The Impact of Lean Production on Operational Performance*.

Matthew Moorhead

Dr. Matthew Moorhead is a property scholar whose work bridges theory and practice, with particular emphasis on development feasibility, property investment and housing affordability. His research interrogates the ways financial modelling, risk management and policy frameworks shape development outcomes and guide investment decisions. Findings have been published in the Journal of Property Investment & Finance, Buildings, and Construction Management and Economics. Cross-disciplinary and international collaborations span ESG integration, early-stage feasibility risk and the uptake of clean-energy technologies in residential property. In addition to supervising higher-degree research candidates, Dr Moorhead works closely with industry partners to ensure that his research remains practically relevant and forward-looking.

Maximilian Neuger

Maximilian Neuger is an Adjunct Lecturer in Property Economics at both the Queensland University of Technology (QUT) and Bond University's School of Built Environment. His research explores the various facets of property as both a tangible and intangible asset within the built environment. For his PhD, Maximilian has worked on government-funded projects examining property market impacts of major infrastructure and is actively engaged in pedagogical innovation, including the application of disruptive technologies in real estate education.

Peter Raptis

Peter Raptis has been a property valuer since 1988 with experience in a wide range of property types. He has a parallel career as a lecturer in property courses at the University of Technology Sydney and the University of New South Wales. He began his teaching career in 1990 and continues to bring practical experience into his classrooms. Peter is a Fellow of the Australian Property Institute. He holds a master's degree in real estate and a second master's degree in property development.

Greg Skulmoski

Associate Professor Dr. Gregory J. Skulmoski has had two career paths: an academic who leads the Project Innovation Management graduate programme at Bond University, and has been a technology project manager. Greg led pioneering technology projects, including cybersecurity, IoT, Lean Six Sigma process improvement, and auditing/compliance projects. Greg used the tools and processes described in *Shields Up*: A *Cybersecurity Project Management* to win the 2017 Middle East Security Award, CISO Council (for risk management in technology projects). He also wrote *Cybersecurity Training: A Pathway to Readiness* (2023) with Chris Walker and *Quantum Cybersecurity Program Management* (2025) with Ashkan Memari, which was an Amazon Top New Release. *Accelerated Quantum Technologies Change Management* (2025 Q4) takes a change management approach to quantum technologies adoption. He is a member of the Tasman Quantum

Network that won the 2025 International Quantum Strategy Day Competition by the Quantum Strategy Institute for developing a quantum business strategy: *From the Edge of the World to the Forefront of Innovation*.

Greg was a core member of the PMBOK® Guide 2000 Update Team that revised the original PMI PMBOK® Guide. He is honoured to have contributed to other standards such as PMI's Practice Standard for Work Breakdown Structures, Organizational Project Management Maturity Model, and Project Manager Competency Development Framework. Dr. Skulmoski is a Certified Information Technology Professional and a lifelong Fellow of the British Computer Society.

Armin Taklif

Armin Taklif is the Founder and Director of AVENTYRA, a specialist consulting firm focused on Digital Engineering (DE), strategy, transformation and management services. With nearly two decades of global experience in the commercial and civil construction industry, Armin shifted his focus in 2015 from managing construction projects to managing information, specialising in DE technologies such as Building Information Modelling (BIM), Geographic Information Systems (GIS), digital twins and related digital systems. He has collaborated closely with Tier 1 consultants, construction contractors and major client organisations across private, local, state and federal government sectors, playing a key role in major DE initiatives on multi-billion-dollar projects. Armin is passionate about exploring innovative technologies, delivering practical digital solutions and driving better outcomes across businesses, projects and asset management.

Adam Wallace

Adam Wallace is an Equity Partner and Specialist Valuer at Australis Asset Advisory Group and is an Associate of the Australian Property Institute (Certified Practising Valuer) and Registered Valuer in Queensland and Western Australia. Since beginning his career in property valuation in 2001, Adam has combined his expertise in Information Technology and programming with his valuation practice, offering a unique and innovative approach to the industry. Now specialising in Insurance, Financial Reporting and Strategic Advisory valuations, Adam is particularly focused on developing and implementing efficient systems and processes through the use of emerging technologies. His dual skill set enables him to enhance valuation methodologies and deliver streamlined, technology-driven solutions for complex valuation challenges.

Elise Wallace

Elise Wallace is an Equity Partner and Specialist Valuer at Australis Asset Advisory Group and is an Associate Member of the Australian Property

Institute (Certified Practising Valuer) and Registered Valuer in Queensland. She brings over 13 years of specialised experience in Public Sector and Not-for-Profit valuations, complemented by 11 years of additional expertise in residential, development site and commercial property valuations, as well as research and data analysis. Elise has a particular interest in strategic advisory and asset management and has also contributed to the professional development of the industry by guest lecturing on Financial Reporting Valuation.

Keith Wolf

Mr. Wolf (SRA, AI-RRS) has been an appraiser and Realtor since 1981. He is a certified residential credentialed appraiser. He earned the RM designation from the American Institute of Real Estate Appraisers and was awarded the SRA and AI-RRS designations from the Appraisal Institute. He was a managing Real Estate broker in seven different states in the United States. Mr. Wolf is a Certified Distance Education Instructor, developing several courses in data analysis. He has a master's degree in data science and is the founder of the LinkedIn group Real Estate Appraiser Data Analytics.

Mr. Wolf managed a $40 billion residential non-performing loan portfolio using alternative valuation methods, including cascading AVMs, from 1995 to 1999. He was part of the team that developed the first automated underwriting system for residential mortgages in 1993. Approximately 50% of his career has been as a Collateral Risk Officer at several national mortgage companies. He has managed credit collateral risk in every regional and national housing crash in the United States from 1981 to the recent out-of-sample increases in prices during the COVID pandemic.

He currently works for a Government Service Entity (GSE) in the US as part of the Credit Collateral Risk Innovation and Modernization team, developing tools that provide transparency in the valuation process, mitigating collateral risk using computer vision, LLMs and various data science methods.

Rita Wong

Rita Wong has around 30 years of experience in the valuation profession, having worked with two of the world's largest listed real estate consultancy companies. Her expertise spans Asia, with a strong focus on Greater China.

Hao Wu

Dr. Hao Wu is a Senior Lecturer in Property at the University of Melbourne. He has conducted research on real estate markets and valuation, property rights and land (re)development in China, Australia and the Asia-Pacific regions.

Chapter 1

Valuation at a crossroads

From resistance to integration of human-digital systems in the profession

Johari Hussein Nassor Amar

The valuation profession and technological context

The real estate sector is at a critical juncture, with unprecedented digital innovations challenging its long-standing professional practice and expertise. Contemporary developments in artificial intelligence (AI), machine learning (ML) algorithms and automated valuation model (AVM) systems extend beyond incremental tool adoption to fundamentally reshape the epistemological foundations of valuation practice. This transformation carries significant implications given that real estate is nearly four times the value (or around 60%) of global wealth (Tostevin & Rushton, 2023), with valuation outcomes influencing mortgage lending, corporate reporting, taxation policy and investment decision-making across multiple economic sectors (Acharya & Ryan, 2016). How the profession's response to technological integration will determine both its future relevance and public confidence in the financial systems dependent upon accurate property assessment.

In October 2021, Zillow abruptly withdrew from automated home buying after accumulating over US$880 million in losses. This marked a watershed moment for the global property valuation industry (Parker, 2022). The failure of algorithmic valuation models of the company to accurately predict acquisition prices exposed fundamental tensions between technological capability and professional judgement. Such tensions continue to define contemporary valuation practice across multiple jurisdictions. This high-profile collapse illustrates a broader challenge confronting valuation professionals worldwide as they navigate the integration of AI and other emerging technologies. Whilst innovative technologies offer great opportunities to enhance accuracy, efficiency and analytical depth, their implementation demands careful consideration of professional standards, regulatory compliance and public trust.

This volume argues that successful AI integration in property valuation requires understanding three interconnected dynamics that shape professional practice. These dynamics include the evolution of professional expertise in technology-augmented environments and the institutional conditions that enable or constrain technological adoption, along with the governance

DOI: 10.1201/9781003539254-1

frameworks necessary to maintain accountability in automated decision-making processes. Rather than viewing AI as either a panacea or a threat to traditional practice, the chapters collectively demonstrate that the future of valuation lies in developing complementarity between human expertise and computational capabilities (Seagraves, 2023). This approach recognises that technological advancement must be balanced with professional judgement and ethical responsibility. The contributors examine how this balance can be achieved whilst maintaining the integrity and reliability that characterise professional valuation practice.

The urgency of this integration challenge extends well beyond individual practice considerations and affects the broader economic system. As AVMs, ML and emerging technologies like LiDAR and drone-based inspection systems become increasingly embedded in valuation processes, fundamental questions arise about professional competency and public accountability. The key question concerns maintaining technical expertise, ensuring methodological transparency and preserving public confidence in valuation outcomes. The response from the professions to technological change will have far-reaching implications for economic stability and market confidence.

Conceptual frameworks for technology-professional integration

Three conceptual frameworks provide the analytical foundation for examining how AI and emerging technologies are transforming property valuation practice. These frameworks move beyond simple adoption models by examining the institutional, social and professional conditions that influence how AI, LiDAR and drone technologies are applied across different valuation contexts. Professional adaptation to technological capability forms the first area of analysis, whilst institutional influences on technology adoption constitute the second, and accountability in automated decision-making represents the third. Each framework provides a distinct analytical perspective on how professional roles and responsibilities are being transformed by technological integration. Together, they establish a foundation for examining both the opportunities and constraints that characterise technology-supported valuation practice, as well as reveal why technological adoption patterns vary considerably across jurisdictions, sectors and property types.

The first framework, *augmented professionalism*, examines how technology can extend but not replace expert judgement in valuation practice. Rather than viewing automated systems as threats to established roles, this approach demonstrates how professionals retain authority by determining how tools are applied in specific contexts (Susskind & Susskind, 2015). The framework aligns with scholarship on connective professionalism, where human and machine contributions are integrated through workflows that preserve professional discretion and interpretive responsibility (Noordegraaf, 2020). In valuation contexts, AI may assist with data modelling or attribute classification, LiDAR enhances measurement precision and drones provide comprehensive visual documentation. However, professional expertise remains essential to assess relevance, apply outputs appropriately and translate technical results into defensible value conclusions. This approach positions valuers as active participants in shaping how technological capabilities are integrated into established professional practice.

The second framework, *socio-technical systems theory*, demonstrates that technology adoption in valuation depends more on institutional environments than on technical capabilities alone (Bijker et al., 1987). This perspective helps explain why identical AI models, drone platforms or LiDAR tools may be embraced in one jurisdiction whilst being rejected in another. Legal frameworks, liability concerns and levels of institutional trust in data infrastructure significantly influence how technologies are implemented across different markets and property types (Geels, 2004). Regulatory environments create varying adoption patterns, with drones restricted over residential areas in some jurisdictions but encouraged for mass appraisal purposes in others. Similarly, LiDAR applications may be accepted for agricultural and industrial sites, whilst remaining uncommon for residential valuations. The variations reflect broader institutional cultures and risk management approaches that shape what constitutes acceptable and defensible professional practice.

The third framework, *algorithmic accountability*, addresses the need for transparency, fairness and responsibility when automated or sensor-based systems support valuation decisions (Kroll et al., 2017). This consideration becomes critical when outputs from AI, drones or LiDAR inform financial reports, insurance assessments or legal proceedings. Professionals must be able to demonstrate how such outputs were generated, what assumptions underpinned them and how they were verified against established standards. When algorithms identify zoning inconsistencies, drones capture structural details or LiDAR systems estimate floor areas, the valuer remains responsible for assessing their reliability and contextual relevance. Without clear validation procedures, transparent methodologies and comprehensive documentation, automated findings may lack credibility in professional and legal contexts. This accountability requirement reinforces the essential role of human oversight throughout the technology-assisted valuation process. Combined, these professional, institutional and algorithmic perspectives provide a comprehensive framework for understanding technology adoption that has been lacking in valuation literature.

From calculators to valuation automation

The integration of AI into property valuation represents the latest iteration of a recognisable pattern of technological adoption that has consistently shaped professional practice. The evolution of valuation practice has followed a trajectory of initial resistance followed by gradual acceptance when new technologies are introduced, from the arrival of handheld electronic calculators in the 1970s to the development of AVMs in the 1990s. Calculators, such as the Texas Instruments TI-30 (1976) and Hewlett-Packard HP-38C (1980), marked a significant departure from traditional techniques that relied exclusively on manual computation and compound interest tables (Taylor, 2010; Tibúrcio Silva, 2023). The tools enabled practitioners to solve complex time value of money problems, calculate net present value and internal rate of return for uneven cash flows, and apply sophisticated financial models that had previously required laborious manual computation (Jalbert & Jalbert, 2019). By the mid-1980s, programmable calculators like the HP-12C had become standard tools in financial and property institutions, despite initial scepticism about their impact on professional judgement and computational rigour (Boer, 1999; Taylor, 2010). This historical pattern provides important context for understanding current responses to AI and emerging technologies in valuation practice.

Similar resistance emerged towards the discounted cash flow (DCF) method, which, although theoretically robust, was initially viewed as too complex and reliant on subjective assumptions regarding discount rates, projected income and holding periods (Crosby et al., 1998; Gabrielli & French, 2021). However, as property markets grew more complex and growth-explicit investment models became more common, traditional valuation methods became inadequate to meet the analytical demands of institutional investors (Trott, 1980). Greater reliance on income-based approaches, notably DCF models in spreadsheet software, became necessary to address requirements for transparency and analytical rigour (Klamer et al., 2018; Gabrielli & French, 2021). Educational institutions incorporated modelling platforms and programmable calculators into curricula (Taylor, 2010), whilst professional bodies such as the Royal Institution of Chartered Surveyors (RICS) and the International Valuation Standards Council (IVSC) published guidance on DCF application and incorporated income-based methods into global standards (Jefferies, 2009). These changes formalised a shift towards more transparent and analytically rigorous practice, enabling valuers to perform scenario testing and sensitivity analysis for properties with complex tenancy arrangements.

The introduction of personal computers and online data platforms in the 1970s and 1980s enabled digital workflows and led to the commercial deployment of AVMs from the late 1980s to the early 1990s (Glumac & Des Rosiers, 2021). Systems progressed into desktop valuation tools during the 1990s and 2000s (Chan & Abidoye, 2019). From the late 2010s to early 2020s, the real estate sector and the broader architecture, engineering and construction industry began embedding computational systems with big data and Internet of Things (IoT) technologies (Börner et al., 2020; Hemmati et al., 2024; Ullah et al., 2018). Corporate real estate valuation has shifted away from traditional inspections and comparable analysis towards data-rich, algorithm-driven methods through technological advancements (Baum et al., 2020; Kok et al., 2017). The current integration of AI and ML represents a continuation of this technological evolution rather than a fundamental departure from established patterns of professional adaptation.

Contemporary technological adoption in valuation has accelerated beyond earlier transitions due to improved efficiency, speed and scale compared to tools from the programmable calculator and DCF era (Baum et al., 2021; Jefferies, 2009; Taylor, 2010). However, the efficiency gains have made valuation processes less transparent than their predecessors. Unlike calculators or spreadsheets, contemporary valuation tools can obscure the link between practitioners and the underlying data that inform value estimates (Valier, 2020). This separation raises concerns about professional accountability and whether outcomes in digital valuation workflows can be properly verified, particularly as AVMs become more widespread. The challenge for the profession lies in maintaining the benefits of technological advancement whilst preserving the transparency and accountability that underpin professional credibility and public trust.

Contemporary challenges and opportunities

The accelerating pace of technological adoption in current market practices across multiple jurisdictions reveals both the opportunities and limitations of AI-enabled valuation systems. In Australia, CoreLogic (2023) reported that over 80% of online lenders employed digital valuations, with AVMs accounting for 28% of assessments and desktop valuations more than doubling between 2019 and 2023. Whilst efficient, the tools are not universally reliable, particularly in markets with atypical properties or sparse data, where automated outputs can diverge significantly from sale prices. Suncorp Bank permits AVMs for properties under $1.5 million when loan-to-value ratios are below 75% but requires additional review if the AVM estimate deviates by more than 5% from the contract price (Suncorp Bank, 2019). This approach reflects the need to balance efficiency gains with accuracy requirements whilst maintaining appropriate risk management protocols – a pattern observed across international markets, where technological adoption is shaped by local conditions and regulations.

The COVID-19 pandemic served as a driver of technological adoption whilst simultaneously exposing the limitations of automated systems in volatile market conditions. UK lenders such as Halifax and Nationwide increased reliance on AVMs to continue mortgage operations amid inspection restrictions imposed by lockdown measures. However, discrepancies between AVM-generated valuations and those provided by independent valuers led to undervaluations that delayed lending for borrowers and constrained financing (The Intermediary, 2024). In response, RICS acknowledged the limitations of automated methods and reaffirmed the need for professional judgement and due diligence (Baum et al., 2021). These experiences highlight the significance of maintaining human oversight in technologically assisted valuations, with the pandemic serving as a natural experiment in balancing digital efficiency and professional judgement.

The collapse of Zillow's iBuying programme in 2021 provides perhaps the most instructive example of the risks associated with algorithmic overconfidence in property valuation. The algorithm of the company persistently overestimated property values, resulting in overpayments and more than US$880 million in resale losses (Parker, 2022). This failure triggered regulatory responses, with the U.S. Consumer Financial Protection Bureau (CFPB) introducing requirements for financial institutions and AVM developers to implement safeguards against algorithmic bias (CFPB, 2022). The Zillow case demonstrates that even sophisticated ML models can fail when market conditions change rapidly or when training data lacks relevant market dynamics. This has informed evolving regulatory standards globally, with authorities increasingly requiring transparency in automated valuations and mandating human oversight for high-value or complex assessments.

International responses to the challenges have varied according to local regulatory cultures and market structures. In Europe, uptake of AI-driven valuation has grown due to speed and cost-effectiveness benefits but has attracted criticism over data quality, model transparency and error rates in markets with outdated or limited data (Topraklı, 2025). The European AVM Alliance and the European Mortgage Federation responded in 2025 with a joint paper urging the use of hybrid valuation models that combine automated outputs with professional oversight (EAA and EMF, 2025). Similar frameworks are gaining ground in Asia and the Middle East, where firms such as ValuStrat now use AI tools in tandem with expert assessors to improve consistency and reduce risk (Sing et al., 2020). These developments demonstrate a growing recognition that effective AI integration depends on technological capabilities with professional expertise, not replacing it through full automation.

The emergence of new technologies such as generative AI, LiDAR scanning and drone-based imaging marks a qualitative shift from earlier technological adoptions in valuation practice. These tools fundamentally alter how evidence is produced and interpreted. For example, AI systems capable of generating textual commentary resembling traditional valuation reports raise concerns about authorship, accountability and professional authority (Alsahan & AlZaidan, 2024; Baum et al., 2021). Meanwhile, LiDAR and drone-based imaging have begun to replace physical site inspections by creating digital property models that demand specialised technical interpretation. This reconfiguration of the role of professional judgement occurs and challenges established norms around direct engagement. As technological processes increasingly intersect with evaluative functions, the boundary between data processing and professional analysis becomes blurred, prompting renewed debate about what constitutes acceptable evidence in valuation processes

(Lee et al., 2022). These advancements call for more sophisticated approaches to education, regulatory oversight and quality assurance than past innovations did, with growing calls for updated ethical codes and expanded guidance to help valuers navigate such complexities (Cheung, 2024).

Volume structure and contributors of the chapters

AI and emerging digital technologies are reshaping property and asset valuation through four interconnected sections that progress from theoretical foundations to practical applications, educational implications and future directions as explored in the volume. The organisation reflects the central argument of the volume that successful technology integration requires understanding the interplay between professional expertise, institutional frameworks and accountability mechanisms. Each section builds upon the three conceptual frameworks established earlier, demonstrating their application across diverse valuation contexts whilst addressing the needs of both academic researchers and professional practitioners.

Section 1: Digital foundations and frameworks in valuation practice

This section establishes the technical, regulatory and professional foundations necessary for understanding AI integration in modern valuation practice. Crawford and Brogan (Chapter 2) examine emerging AI technologies against professional standards and legal compliance requirements. Their analysis demonstrates how the augmented professionalism framework enables valuers to maintain authority whilst using technological capabilities. Khanjanasthiti, Gardner and Taklif (Chapter 3) analyse AI and ML integration within government valuation services. They illustrate how socio-technical systems theory explains varying adoption patterns across institutional contexts. Wu and Wong (Chapter 4) explore coevolutionary frameworks for understanding mutual adaptation between AI technologies and the valuation profession. Their work challenges linear disruption models and highlights the importance of human-AI partnerships. Ge and Raptis (Chapter 5) investigate data-related challenges in applying AI to real estate valuation. They highlight the importance of an algorithmic accountability framework for responsible AI deployment through robust data strategies and human oversight.

Section 2: Applied digital innovations in valuation practice

This section demonstrates how theoretical possibilities translate into operational realities across diverse professional contexts. Wallace, Wallace, Moorhead and Logue (Chapter 6) examine emerging technology adoption in public sector valuations. They show how institutional factors influence the acceptance and resistance patterns predicted by socio-technical systems theory. Khanjanasthiti, Taklif and Gardner (Chapter 7) explore the transition from conventional to drone-based airport runway inspections. Their case study demonstrates how augmented professionalism manifests in specialised infrastructure assessment contexts. Wolf and Amar (Chapter 8) introduce generative AI prompt engineering fundamentals for real estate appraisers. They provide practical guidance that embodies the algorithmic accountability framework by maintaining professional standards and human oversight in AI-assisted practice. These chapters collectively illustrate how the conceptual frameworks of the volume operate in real-world implementation scenarios.

Section 3: Innovating valuation education and training for the digital Era

This section addresses the challenge of preparing professionals for an AI-enhanced valuation landscape whilst preserving core competencies and ethical standards. Boyd (Chapter 9) examines new pedagogical frameworks for integrating AI into real estate education through the Apply, Evaluate and Solve (AES) model. This chapter shows how educational institutions can address the socio-technical challenges of preparing professionals for technology-augmented practice. Neuger and Armitage (Chapter 10) investigate spatial computing's transformative potential in valuation education through Apple Vision Pro and augmented reality technologies. They demonstrate how augmented professionalism principles can be embedded in educational design. The section contributions reveal how educational approaches must evolve to support the professional adaptation required by the analytical framework of the volume.

Section 4: Reflections on legal, ethical and future directions in digital valuation

This section examines broader implications of AI integration for legal accountability, risk management and the future trajectory of valuation practice. Cradduck (Chapter 11) provides a comprehensive legal analysis of generative AI use in valuation practice. This chapter directly addresses the algorithmic accountability framework's requirements for professional liability and procedural compliance. Fonteyn (Chapter 12) reflects on current AI, big data and analytics integration through industry expert perspectives. The chapter synthesises insights into the evolving balance between automated systems and human professional judgement. Skulmoski, Lang, Memari and Lee (Chapter 13) explore the transformational potential of quantum computing for property valuation, extending the analysis of the volume to emerging technologies that may further reshape professional practice. Overall, these chapters demonstrate how the conceptual frameworks of the volume provide analytical tools for understanding both current challenges and future developments.

Intended audiences and practical applications

Multiple interconnected audiences within the property valuation ecosystem, each requiring different but complementary perspectives on AI integration challenges, are served by the volume. The analysis enables readers to move from theoretical understanding to practical implementation of AI integration strategies. For academic researchers, the volume provides theoretical frameworks for analysing technology-profession relationships, empirical evidence from diverse jurisdictions and methodological approaches for studying professional adaptation to technological change (Susskind & Susskind, 2015; Noordegraaf, 2020). The three conceptual frameworks offer analytical tools for understanding how professions maintain authority and accountability whilst integrating advanced technologies, contributing to broader scholarship on professional work, innovation studies and science and technology studies (Bijker et al., 1987; Geels, 2004). By combining an interdisciplinary approach with theoretical frameworks, the volume advances understanding of how professional domains adapt to technological disruption, offering insights to wider debates on automation, expertise and governance.

For valuation practitioners, the volume offers practical guidance on AI adoption whilst maintaining professional standards and regulatory compliance (IVSC, 2024). Case studies and implementation examples from Australian, UK and USA contexts provide useful insights into effective technology integration strategies, risk management and quality assurance (CoreLogic, 2023; Suncorp Bank, 2019). Practitioners will find value in the chapters addressing legal liability, ethical concerns and regulatory obligations, which translate theoretical ideas into actionable strategies for practice development (Baum et al., 2021; CFPB, 2022). The volume's focus on augmented professionalism offers a clear framework for sustaining professional authority and accountability whilst leveraging algorithmic tools to improve service delivery and analytical depth. This practical orientation not only supports immediate implementation but also contributes to long-term professional development in an increasingly digital work environment (Chan & Abidoye, 2019).

Educational institutions and trainers will benefit from the volume's analysis of pedagogical strategies for preparing professionals for technology-augmented practice (Taylor, 2010). Chapters on educational innovation and professional development provide practical models for curriculum design, teaching methods and skill-building programmes that integrate technological competency with professional judgement. The volume highlights the need for evolving educational approaches that reflect practice environments where human expertise and AI are increasingly interconnected (Wheaton & Xu, 2024). This guidance is vital for upholding professional standards and ensuring graduates are equipped for the realities of contemporary valuation work. The educational frameworks presented support both immediate curriculum development needs and longer-term strategic planning for professional preparation programmes.

Regulatory authorities and policymakers will find the volume's exploration of governance frameworks, accountability mechanisms and international comparison approaches valuable for shaping oversight structures for AI-enabled valuation practice (Kroll et al., 2017). By examining diverse regulatory responses to technological adoption, it presents insights into policy approaches that foster innovation without compromising public trust and professional accountability (Topraklı, 2025). It emphasises transparency, fairness and responsibility in automated decision-making and directly engages with concerns about algorithmic governance and professional oversight. Drawing on international case studies, the volume provides evidence-based guidance for developing regulatory strategies that balance technological advancement with consumer protection and market stability requirements (The Intermediary, 2024).

The volume's practical relevance extends to technology developers and vendors aiming to understand the professional requirements for AI tools in valuation contexts (Sing et al., 2020). Its analysis of both successful and unsuccessful technology implementations delivers guidance for designing tools that align with established workflows, regulatory expectations and ethical standards (Parker, 2022). The three conceptual frameworks help technology developers create systems that complement rather than replace professional judgement, reinforcing accountability and supporting sustainable practices that meet professional needs, respect regulatory boundaries and adhere to ethical obligations, which will likely achieve long-term adoption. The technical insights throughout the volume promote collaboration between technology developers and practitioners in creating effective AI-driven valuation solutions.

This volume equips its diverse audiences with the critical perspectives needed to navigate AI integration of emerging technologies into property valuation practice. The conceptual frameworks and empirical insights form a foundation for understanding how professional expertise and technological tools can work in concert to preserve accountability and enhance analytical effectiveness. The profession cannot afford to stand still because technology is ongoing and uneven, but the evidence gathered here demonstrates that thoughtful integration is both possible and necessary. How the profession responds over the next decade will determine whether valuers continue to play a central role in the property sector or become marginalised by systems they neither shaped nor governed.

Discussion questions

1 How does the historical pattern of resistance and acceptance, i.e., from calculators to DCF models to AVMs, inform current responses to AI in valuation practice?
2 How did COVID-19 expose both the potential and the limits of technology-enabled valuation, and what does this reveal about the evolving role of technology in professional judgement and practice?

3 How can valuation practitioners adopt AI effectively in the international market whilst balancing competitive advantage, regulatory compliance, transparency and accountability?

Further reading

International Valuation Standards Council (IVSC). (2025, July 23). Navigating the rise of AI in valuation: Opportunities, risks, and standards (Perspectives Paper). *IVSC*. Available online: https://ivsc.org/navigating-the-rise-of-ai-in-valuation-opportunities-risks-and-standards/ (Accessed 20 July 2025)

Royal Institution of Chartered Surveyors (RICS). (2022, April). Automated valuation models (AVMs): Implications for the profession and their clients (Insight Paper). *RICS*. Available online: https://www.rics.org/profession-standards/rics-standards-and-guidance/sector-standards/valuation-standards/automated-valuation-models (Accessed 20 July 2025)

References

Acharya, V. V., & Ryan, S. G. (2016). Banks' financial reporting and financial system stability. *Journal of Accounting Research*, *54*(2), 277–340. https://doi.org/10.1111/1475-679X.12114

Alsahan, I. M., & AlZaidan, Z. I. (2024). Unleashing the power of artificial intelligence in real estate valuation: Opportunities and challenges ahead. *Journal of Artificial Intelligence General science (JAIGS)*, *3*(1), 89–96. ISSN: 3006-4023. https://ojs.boulibrary.com/index.php/JAIGS/article/view/69

Baum, A., Graham, L., & Xiong, Q. (2021). *The Future of Automated Real Estate Valuations (AVMs)*. Oxford: University of Oxford. Retrieved on 24 May 2025 from https://www.sbs.ox.ac.uk/sites/default/files/2022-03/FoRE%20AVM%202022.pdf

Baum, A., Saull, A., & Braesemann, F. (2020). *PropTech 2020: The Future of Real Estate*. University of Oxford. Retrieved on 24 May 2025 from https://www.sbs.ox.ac.uk/sites/default/files/2020-02/proptech2020.pdf

Bijker, W. E., Hughes, T. P., & Pinch, T. J. (eds.) (1987). *The Social Constructions of Technological Systems: New Directions in the Sociology and History of Technology*. Cambridge, MA: MIT Press.

Boer, F. P. (1999). *The Valuation of Technology—Business and Financial Issues in R&D*. New York: John Wiley & Sons, Inc.

Börner, K., Scrivner, O., Cross, L. E., Gallant, M., Ma, S., Martin, A. S., Record, L., Yang, H., & Dilger, J. M. (2020). Mapping the co-evolution of artificial intelligence, robotics, and the internet of things over 20 years (1998-2017). *PLoS One*, *15*(12), e0242984. https://doi.org/10.1371/journal.pone.0242984

CFPB (2022). Consumer financial protection bureau outlines options to prevent algorithmic bias in home valuations. Retrieved on 24 May 2025 from https://www.consumerfinance.gov/about-us/newsroom/cfpb-outlines-options-to-prevent-algorithmic-bias-in-home-valuations/

Chan, A. P., & Abidoye, R. B. (2019). Advanced property valuation techniques and valuation accuracy: Deciphering the artificial neural network technique. *RELAND: International Journal of Real Estate & Land Planning*, *2*, 1–9. https://doi.org/10.26262/reland.v2i0.6748

Cheung, K. S. (2024). Real estate insights: Establishing transparency–setting AI standards in property valuation. *Journal of Property Investment & Finance*, *42*(4), 406–408. https://doi.org/10.1108/JPIF-04-2024-0050

CoreLogic. (2023, April 27). Digital valuations surge, as data transforms the mortgage industry. Retrieved on 24 May 2025 from https://www.corelogic.com.au/news-research/news/2023/digital-valuations-surge%2C-as-data-transforms-the-mortgage-industry

Crosby, N., Lavers, A., & Murdoch, J. (1998). Property valuation variation and the 'margin of error' in the UK. *Journal of Property Research*, *15*(4), 305–330. https://doi.org/10.1080/095999198368310

EAA and EMF (2025). Joint paper on the use of automated valuation models in Europe. Retrieved on 24 May 2025 from https://www.europeanavmalliance.org/files/eaa/Downloads/Archive/EMF_EAA_Survey_on_AVMs_in_Europe.pdf

Gabrielli, L., & French, N. (2021). Pricing to market: Property valuation methods–A practical review. *Journal of Property Investment & Finance*, *39*(5), 464–480. https://doi.org/10.1108/JPIF-09-2020-0101

Geels, F. W. (2004). From sectoral systems of innovation to socio-technical systems: Insights about dynamics and change from sociology and institutional theory. *Research Policy*, *33*(6–7), 897–920. https://doi.org/10.1016/j.respol.2004.01.015

Glumac, B., & Des Rosiers, F. (2021). Towards a taxonomy for real estate and land automated valuation systems. *Journal of Property Investment & Finance*, *39*(5), 450–463. https://doi.org/10.1108/JPIF-07-2020-0087

Hemmati, A., Raoufi, P., & Rahmani, A. M. (2024). Edge artificial intelligence for big data: A systematic review. *Neural Computing and Applications*, *36*(19), 11461–11494. https://doi.org/10.1007/s00521-024-09723-w

IVSC (2024). Inspections of tangible assets as part of the valuation process: IVSC perspectives paper. Retrieved on 24 May 2025 from https://www.ivsc.org/wp-content/uploads/2024/06/IVSC-Perspectives-Paper-Inspections.pdf

Jalbert, T., & Jalbert, M. (2019). A comparison of financial calculators. *Journal of Financial Education*, *45*(1), 58–71. https://www.jstor.org/stable/26918024

Jefferies, R. L. (2009). A brief history and development of 'real value' valuation models – The last four decades. *Paper presented at the Pacific Rim Real Estate Society Conference*, 18–21 January, Sydney, Australia. Retrieved on 24 May 2025 from http://www.prres.net/papers/Jefferies_A_Brief_History_And_Development_Of.pdf

Klamer, P., Bakker, C., & Gruis, V. (2018). Complexity in valuation practice: An inquiry into valuers' perceptions of task complexity in the Dutch real estate market. *Journal of Property Research*, *35*(3), 209–233. https://doi.org/10.1080/09599916.2018.1510429

Kok, N., Koponen, E. L., & Martínez-Barbosa, C. A. (2017). Big data in real estate? From manual appraisal to automated valuation. *The Journal of Portfolio Management*, *43*(6), 202–211. https://doi.org/10.3905/jpm.2017.43.6.202

Kroll, J. A., Huey, J., Barocas, S., Felten, E. W., Reidenberg, J. R., Robinson, D. G., et al. (2017). Accountable Algorithms. *165 U. Pa. L. Rev*. 633. Available online at: https://scholarship.law.upenn.edu/penn_law_review/vol165/iss3/3

Lee, C. C., Lee, H. Y., Yeh, W. C., & Yu, Z. (2022). The impacts of task complexity, overconfidence, confirmation bias, customer influence, and anchoring on variations in real estate valuations. *International Journal of Strategic Property Management*, *26*(2), 141–155. https://doi.org/10.3846/ijspm.2022.16704

Noordegraaf, M. (2020). Protective or connective professionalism? How connected professionals can (still) act as autonomous and authoritative experts. *Journal of Professions and Organization*, *7*(2), 205–223. https://doi.org/10.1093/jpo/joaa011

Parker, W. (2022, February 10). Zillow's shuttered home-flipping business lost $881 million in 2021. *The Wall Street Journal*. Retrieved on 24 May 2025 https://www.wsj.com/business/earnings/zillows-shuttered-home-flipping-business-lost-881-million-in-2021-11644529656

Seagraves, P. (2023). Real estate insights: Is the AI revolution a real estate boon or bane?. *Journal of Property Investment & Finance*, *42*(2), 190–199. https://doi.org/10.1108/JPIF-05-2023-0045

Sing, T., Yang, J., & Yu, S.-M. (2020). Decision tree and boosting techniques in artificial intelligence based automated valuation models (AI-AVM). *SSRN Electronic Journal 3605798*, 1–36.

Suncorp Bank. (2019, October 31). 5.34 - Valuations: Policy and underwriting standards (Version 38): Submission to the House of Representatives Standing Committee on Economics, Banks and financial institutions. Retrieved on 24 May 2025 from https://www.aph.gov.au/-/media/02_Parliamentary_Business/24_Committees/243_Reps_Committees/Economics/46p/Banks_and_Financial_Institutions/3_Smaller_Banks/Suncorp/SCB031QW.pdf

Susskind, R., & Susskind, D. (2015). *The Future of the Professions: How Technology will Transform the Work of Human Experts*. London: Oxford University Press.

Taylor, R. W. (2010). Business finance and financial calculators: A historical and pedagogical perspective. *Journal of Business Administration Online*, *9*(2), 1–9.

The Intermediary (2024, February 22). Brokers report a rise in down valuations. Retrieved on 24 May 2025 from https://theintermediary.co.uk/2024/02/brokers-report-a-rise-in-down-valuations/

Tibúrcio Silva, C. A. (2023). Discounted cash flow: Theoretical and practical aspects of its use. *Advances in Scientific & Applied Accounting*, *16*(3). https://doi.org/10.14392/asaa.2023160309

Topraklı, A. Y. (2025). AI-driven valuation: A new era for real estate appraisal. *Journal of European Real Estate Research*, *18*(1), 105–120. https://doi.org/10.1108/JERER-04-2024-0031

Tostevin, P., & Rushton, C. (2023). Total value of global real estate: Property remains the world's biggest store of wealth. Retrieved on 24 May 2025 from https://impacts.savills.com/market-trends/the-total-value-of-global-real-estate-property-remains-the-worlds-biggest-store-of-wealth.html

Trott, A. (1980). *RICS Research Project into Property Valuation Methods: Interim Report*. London: Polytechnic of the South Bank.

Ullah, F., Sepasgozar, S. M., & Wang, C. (2018). A systematic review of smart real estate technology: Drivers of, and barriers to, the use of digital disruptive technologies and online platforms. *Sustainability*, *10*(9), 3142. https://doi.org/10.3390/su10093142

Valier, A. (2020). Who performs better? AVMs vs hedonic models. *Journal of Property Investment & Finance*, *38*(3), 213–225. https://doi.org/10.1108/JPIF-12-2019-0157

Wheaton, W. C., & Xu, C. (2024). *Using AI to Improve Price Transparency in Real Estate Valuation*. MIT Center for Real Estate Research Paper No. 24/16. Available at SSRN: http://doi.org/10.2139/ssrn.5056315

Section 1

Digital foundations and frameworks in valuation practice

Chapter 2

Emerging AI technologies in the valuation of tangible assets

Allen J Crawford and Kevin Brogan

Introduction

Artificial Intelligence (AI) in the property industry commenced well before the 1980s. Over the last 60 years, there has been a continual evolution of new technologies. This chapter:

- i initially puts AI in context;
- ii sets the valuation framework within which AI must fit;
- iii outlines the importance of transparency in the valuation process;
- iv reviews the history of AI in the property industry;
- v discusses AI in a legal context;
- vi analyses how AI is useful for statutory valuation purposes;
- vii considers the viewpoints contrary to those of the authors of this chapter;
- viii forms a conclusion on 'Emerging AI Technologies in the Valuation of Tangible Assets'.

This paper demonstrates that artificial intelligence in any of its forms is just one of the many tools that valuers can use in the valuation of tangible assets. No tool or combination of tools can replace the fact that value is a '*Valuer's opinion of value, in monetary terms, for a tangible asset as a result of a valuation process as at a specified date*' (Australian Property Institute, 2024). That is, if there is no valuer, then the result is not a valuation.

Artificial intelligence concepts

The Oxford Dictionary defines:

artificial: *made or produced to copy something natural; not real.*
intelligence: *The faculty of understanding; intellect*
Artificial Intelligence: *The capacity of computers or other machines to exhibit or simulate intelligent behaviour*

DOI: 10.1201/9781003539254-3

There are two different forms of AI: Traditional and Generative (Wohlfarth 2024). The differences between AI and Generative AI are as follows:

> Traditional AI (also referred to as Narrow or Weak AI) is a subset of artificial intelligence that is used to perform tasks based on predetermined algorithms. Usually, traditional AI-based solutions excel at carrying out a limited set of tasks, being more restricted in the scope of potential applications than other types of AI.
>
> Generative AI is an artificial intelligence that is capable of generating outputs such as text, images, and other data. This type of AI works primarily by taking large quantities of existing data, analysing it, and based on those findings, producing new content. What this means is that generative AI relies on **machine learning** to recognise, predict, and create content using the data sets it can access.
>
> Generative AI (Gen AI) is a form of artificial intelligence that is capable of creating new content, including text, images or sounds, based on patterns and data acquired from a body of training material. That training material may include information obtained from 'scraping' publicly and privately available text sources to produce large language models.
>
> (Chief Justice New South Wales 2024)

The valuation process

When considering the use of emerging AI technology, valuation professionals must remain compliant with professional valuation, ethical and regulatory standards whilst understanding and adapting to technological advancements.

Trust and confidence in valuations are critical to maintaining a stable economy. The overarching concepts and principles of valuation are contained in the International Valuation Standards (IVS), whilst Valuation Professional Organisations (VPOs) (such as the Australian Property Institute API) develop, promote, and implement valuation standards applicable in the relevant jurisdiction as well as being responsible for the regulation of their valuer members. Other bodies such as the Australian Prudential Regulation Authority (APRA), the Valuers Registration Board of Queensland (VRBQ) and the Consumer Protection Division of the Western Australian Department of Local Government, Industry Regulation and Safety provide regulatory oversight for the banking sector and consumer protection, respectively.

The API defines *valuation* and the *valuation process* as follows:

> **'valuation'** means an established evidence-based valuation process for assessing the value, including but not limited to market value and market rent, of a tangible asset as at a specified date following a physical inspection of the asset by the Primary Valuer.
>
> (Rules of Professional Conduct - Australian Property Institute, 2024)

> **'valuation process'** means all the necessary enquiries, investigations, procedures and processes including the physical inspection of the tangible asset by the Primary Valuer, which is a pre-requisite for a valuation, required to fully inform the Valuer's reasoning and analysis in accordance with the practice accepted as proper by the API.
>
> (Rules of Professional Conduct - Australian Property Institute, 2024)

A qualified opinion of value can be arrived at by the *desktop process*, which is defined as:

> an evidence-based process which results in a desktop report containing an indicative value of a tangible asset as at a specified date, based on the Valuers knowledge, research, experience and expertise together with 'relevant data and inputs' (information) relating to the asset provided by other parties or sourced by the Valuer, but does not include a physical inspection (of any kind) of the asset by the Valuer.
>
> (Rules of Professional Conduct - Australian Property Institute, 2024)

Common to each of the above is the involvement of a valuer. The API refers to a valuer as 'an individual possessing the necessary training, qualifications, knowledge, expertise and experience to undertake a valuation process in an objective, unbiased, ethical and competent manner'.

The act or process of forming a conclusion on a value according to IVS requires professional judgement, which is defined as:

> The use of accumulated knowledge and experience, as well as critical reasoning to make an informed decision.

A valuation also requires the application of professional scepticism, which is defined as:

> An attitude that includes a questioning mind and critical assessment of valuation evidence.
>
> (International Valuation Standards Council, 2024)

The IVS Council undertook an agenda consultation during 2024. One of the areas for consultation was the use of technology in valuation. The response from the API stressed that

> valuation is the process the valuer undertakes when providing an IVS-compliant valuation report, and the importance of the valuer's professional judgement and scepticism when it comes to reliance on information and data from others. The valuer is ultimately responsible for the investigations, analysis and conclusions reached as well as the content of the valuation report, including the value.

A sample of other responses included:

- An Automated Valuation Model (AVM) cannot produce an IVS-compliant valuation.
- Valuation must always include professional judgement throughout the process.
- AI is a useful tool for valuers to use in the valuation process.
- AI cannot provide IVS-compliant valuations.
- AI can be used as an aid for the writing of valuation reports.
- AI is developing at such a rate that it is impossible to predict the standard setting.
- AI threatens the valuation profession.

A valuation model is defined in IVS as 'a quantitative implementation of a method in whole or in part that converts inputs into outputs used in the development of a value'. Whenever a valuer provides an opinion of value incorporating valuation approach(es) and method(s), they have technically applied a 'valuation model'.

There is a clear distinction from an Automated Valuation Model (AVM), which is a type of model that provides an automated calculation for a specified *asset* at a specified date, using an algorithm or other calculation technique without the valuer applying professional judgement over the model, including assessing and selecting inputs or reviewing outputs.

The role of data sources is arguably a form of AI. The valuer, however, must undertake necessary enquiry to be satisfied on the correctness and completeness of information on which reliance is placed. This is known as fact-checking and is a critical part of the valuation process.

To put this in a cumulative scenario, unquestioning reliance on data or technology is likely to result in an impact on a valuation that:

- is not the subject of professional scepticism;
- is not the actual opinion of the Primary Valuer;
- allows some other input outside the direct control of the Primary Valuer; and
- cannot be verified by the Primary Valuer.

In such a case, the Professional Conduct Committee of the API would probably arrive at a prima facie decision that the Primary Valuer was guilty of professional misconduct should a complaint be lodged against a valuation conducted by that Primary Valuer. In addition, such a valuation will be difficult to defend if challenged (see the Weber matter discussed under the heading Application in a Legal Environment).

Whilst the above discussion may seem to totally reject AI:

> It is important to note that this is not a call to reject technology, automation and AI outright. Rather, it is a plea for a balanced approach that recognises the immense potential of technological advancements while remaining vigilant against the dangers of over-reliance (driven by some very convincing sales pitches from vendors).
>
> (Bukowski-Kruszyna 2024)

Transparency

Transparency of real estate markets and valuation practice is a cornerstone of trust in the valuation profession and the regulatory framework for valuers. Significant work is being undertaken globally to underline this with organisations such as the G20 and the United Nations.

The V20 Summit and Conference was established to promote the critical role of trusted, reliable, transparent and objective valuation advice in promoting good decision-making, sustainable development and a stable global economy. The V20 Summit and Conference run alongside the G20 and align with the G20 principles of sustainable development, inclusive growth and responsible governance. V20 Summits have been held in India and Brazil, with future summits planned for South Africa and the USA in 2025 and 2026.

The inaugural V20 Summit declaration was signed by international Valuation Professional Organisations (including the API) and includes commitments to continue:

- championing trusted, reliable and transparent advice to foster trust, confidence and prosperity.
- upholding unwavering ethics, striving for methodological excellence and committing to transparent practices.

Transparency is rightly identified as a key pillar to creating trust and confidence in valuation, which is underpinned by the focus in IVS on approaches and methods.

The Federation Internationale des Geometres (International Federation of Surveyors, or FIG) has established a number of commissions to investigate different areas of surveying practice. The mission statement of FIG Commission 9 (Valuation and Management of Real Estate) is:

> FIG takes the lead in the discussion on transparency in real estate markets. The vision is that by 2026 this issue will be on top of the agendas of organisations working globally on issues such as valuation, land management and climate change topics and others.

Commission 9 has three main working groups, and working group 9.1 (Definition of Transparency in Real Estate Markets) is investigating the definition and measurement of transparency in real estate markets. The contention is that 'a fundamental pillar of democracy and fair action is transparency in decision making and action'. The working group is attempting to quantify and report on the level of transparency in national property markets around the world.

FIG Working Group 9.3 (Automated Valuation Modelling (AVM) with machine learning and artificial intelligence) is investigating the impact of AI on the transparency of valuations and has included the World Bank and VPOs in its considerations. The digital age is making new tools and platforms available that can assist the valuer in the valuation process, but this has also raised significant concerns about whether they can be relied upon as part of a valuation. This is particularly the case where the technology is not transparent or understood by the valuer, or if the valuer has not inspected the property.

One question that needs to be asked is, 'who benefits from a lack of transparency?' It is certainly possible that the benefit of opacity will accrue to a data company that has, for instance, a proprietary AVM.

The Asian Development Bank supports the valuation of property. The Bank uses valuations for the acquisition of land for developments and, on that basis, believes that it is unethical to use a method that does not involve an inspection of the property and the professional judgement of a valuer (Assessors and Registered Valuers Foundation, 2023).

For similar reasons, the World Bank Land Acquisition and Rehabilitation Resettlement Policy Framework requires the valuation to be determined by a licensed independent appraiser using recognised International Valuation Standards (Policy 5 of the Framework).

History of AI in the property industry

Where did AI start?

Jonathan Swift envisaged a form of AI when he published Gulliver's Travels in 1726 and described how Gulliver toured the grand academy of Lagadu in Laputa, where he met a professor who demonstrated his 'engine', which would facilitate:

> …by his contrivance, the most ignorant person, at a reasonable charge, and with a little bodily labour, might write books in philosophy, poetry, politics, laws, mathematics, and theology, without the least assistance from genius or study.

In the authors' opinion, AI in the property industry actually commenced well before 1980. Adding machines and calculators are a primitive form of AI.

The adding machine and calculators were an advance on the tool we originally used, called a slide rule. 'Artificial Intelligence has often been loosely used in many different contexts since 1960'.

Conventional wisdom is, however, that in regard to the property of the property, AI can be divided into three stages. These stages broadly correspond to the period from 1980 to 2000, from 2000 to 2008, and from 2008 to the present (Jones & Trevillion 2022) and Reed (2021).

1980 to 2000

Digital technology began to be adopted by the property industry during the 1980s, when personal computers became more common. Spreadsheet and accounting software such as Lotus 1–23 and Microsoft Excel began to be used by real estate companies and valuation professionals. Advancements in software packages to aid investment analysis also allowed real estate investors to more accurately assess the performance value of investment property.

2000 to 2008

Real estate technology first targeted consumers during the dot-com bubble. Marketing of real estate changed from advertising in print media and hard copy distribution of marketing brochures to a primary focus on listings in digital media (Voices 2019). Whilst this occurred, there is still ample evidence of hard copy marketing, particularly residential housing using letterbox drops.

2008 to date

Widespread availability of high-speed internet meant that real estate companies could move more of their data and services online (Reed 2021). Real estate databases, such as Zillow are an example of information such as geographic data, property valuation and real estate advice, being moved online (Soules 2021). This has been successful, with companies like Zillow (US), Rightmove (UK) and Price Hubble (CH) being in the top listed companies in their respective markets.

Recent acceleration in the use of technology in valuation

COVID-19 accelerated the adoption of information technology in the real estate industry (Forlee 2022; Rishi 2021).

Some technology that was adopted 'at pace' during the pandemic was extremely successful, with teleconferencing being an established part of our lives. There are a number of teleconference platforms, such as Skype, Zoom and particularly Teams, that now form a normal part of the life of a property professional. Unfortunately, the rushed adoption of some technology solutions diluted professional and regulatory standards and has had a lasting, detrimental impact.

That is not to say that the technology cannot be adapted and refined to ensure compliance with ethical standards and regulatory requirements, but the lack of proper testing led to some questionable practices and shortcuts that are yet to be resolved to reinstate professional integrity and protect the public interest. For example, property valuation firms purporting desktop assessments are the equivalent of the valuation process where the Primary Valuer inspects the subject property (and the evidence on which reliance is placed).

Nonetheless, the pandemic helped to drive e-commerce and resulted in the closure of many traditional retail stores, which has impacted the commercial real estate industry (Tiwari & Miao 2022). Blockchain technology has also been used to track property for the purposes of land registration and resolve potential ownership disputes (Jone & Trevillion 2022).

Application in a legal environment

Courts rely on the opinion of experts. An expert means a person who would, if called as a witness in the proceeding, be qualified to give opinion evidence in relation to an issue. The expert will inevitably have their opinion tested against the opinion of another expert.

There, however, have been examples where the expert's opinion was discredited due to reliance on AI.

In the matter of Weber vs Weber in the Surrogates Court of Saratoga County (New York), a supposed expert witness relied on Microsoft Copilot, a large language model generative artificial intelligence chatbot, in cross-checking his calculations. Despite his reliance on artificial intelligence, Mr Ranson could not recall what input or prompt he used to assist him with the Supplemental Damages Report. He also could not state what sources Copilot relied upon and could not explain any details about how Copilot works or how it arrives at a given output.

> Mr. Ranson was adamant in his testimony that the use of Copilot or other artificial intelligence tools, for drafting expert reports is generally accepted in the field of fiduciary services and represents the future of analysis of fiduciary decisions; however, he could not name any publications regarding its use or any other sources to confirm that it is a generally accepted methodology.

The court held, 'The use of artificial intelligence is a rapidly growing reality across many industries. The mere fact that artificial intelligence has played a role, which continues to expand in our everyday lives, does not make the results generated by artificial intelligence admissible in Court'.

'The Queensland Law Society (QLS) does not seek to discourage use of AI tools in legal practice, nor does QLS view such use as incompatible with ethical duties', AI – used properly – has the potential to significantly enhance access to justice (Chien and Kim 2024) and lessen the burden of routine work on solicitors and support staff. Balanced against this is the need to ensure that innovation does not lead to negligence or professional censure. Appropriate use of AI requires guardrails and a realistic understanding of the limitations of the product.

QLS continue

> Given the varied needs of individuals and small business, it is highly unlikely that all their legal work can be supplied by hybrid or automated services, no matter how good such systems become. However, to disrupt an industry you do not need to take all the work … Automation works best in matters which are straightforward, predictable, and can be broken into replicable workflows. That kind of matter is also the bread and butter for many SME law firms.

Whether truth or fiction, it is not unreasonable to assume the accuracy of the rumour that a lawyer let AI write closing arguments, including citing precedent and the AI tool invented case law, as it did not have any relevant case law in its database.

The further concern of Generative AI is that it is limited to the technology input, which includes and attempts to understand both the rational and irrational decisions made by humans, relying solely on the decision without understanding the market participant's reason for that decision, which can only be derived by personal discussion with that market participant.

The words of the Chief Justice of New South Wales in the Supreme Court Practice Note SC Gen 23 sum up the use of Gen AI.

> Gen AI should not be used for editing or proofing draft judgements, and not part of a draft judgement should be submitted to a Gen AI programme...
>
> The product of all Gen AI generated research, even if apparently polished and convincing, should be closely and carefully scrutinised and verified for accuracy, completeness, currency and suitability before making any use of it. Gen AI research should not be used as a substitute for personal research by traditional methods…
>
> Judges may require litigants (including litigants in person) and legal representatives including counsel to disclose any use of Gen AI in respect of written submissions or other documents placed before the Court, and may also require an assurance that any such documents have been verified for accuracy, including an identification of the process of verification followed including, where applicable, for the purpose of ensuring compliance with Practice Note SC Gen 23.

In a follow-up to the practice note, Justice Bell said on 4 December 2024,

> From the start of 2025's first law term, legal practitioners caught using generative artificial intelligence (GenAI) in affidavits, witness statements, and other evidentiary documents could be held in contempt of court and will have the material tossed out ... most concerned with ensuring that 'hallucinations' – being the generation of apparently plausible but inaccurate references – are kept out of courtrooms ...'Judges are entitled to expect counsel and solicitor advocates to only present cases [that] are relevant and in fact exist, so we're making no apologies for that, and I don't think anyone should think we should.

Statutory valuations

Statutory valuations are completed in accordance with enabling legislation in the relevant jurisdiction. For the purpose of this discussion, statutory valuations are assessments made for rating and taxation purposes and are made to set the rateable value of the assets, which are used to calculate rates and other property taxes. They are initially mass appraisals in which algorithms are able to be written with valuation experts interpreting the sales evidence into metric categories that then enable those metrics to predict changes in value from the previously issued statutory valuation. An additional consideration for mass appraisal is to ensure consistency in the 'valuation list', and this may override the accuracy of an assessment for an individual property.

The database on which statutory valuations are based can and should be refined and improved with healthy scepticism and proper analysis of evidence by those charged with managing the 'valuation list'. Proper direction and updating of the database by expert professional valuers on the selection of data as inputs to the valuation model (algorithm) may result in more reliable and equitable results.

It is therefore not unreasonable to envisage the extensive use of AI in the short to medium term in statutory valuations, particularly where it is simply looking at land values and not improvements or capital values. This practice is already in place at present, with selected benchmark properties being valued and the metrics derived then being applied to a broader group of properties. It is expected that this process is likely to develop and improve significantly in the future as data sets become more accurate.

In order for the system to work, the database of detailed, analysed sales evidence by professional expert valuers will need to be maintained and improved.

Valuers will need to be directly involved in scenarios where property holders contest or challenge the statutory valuation, and particularly where any matter proceeds to an appeal level where a Court or tribunal requires the input/testimony of a valuation expert.

Shortcomings of AI

An AI system is not able to independently satisfy itself that there is sufficient or appropriate information in the database for the output value of a particular property. AI systems are limited to the database on which they base their conclusions. Simply put, the GIGO (garbage in, garbage out) rule applies.

Xiong (2024) explains why it is so hard (for AI) to estimate a function $y=f(x)$. He explains that where there is a lack of high-quality data around y and x to train the model, which is statistical, machine learning, AI or even basic experience-based guesses.

'The market evolves constantly over time. Figure it out and you are already late'. There is no one-size-fits-all solution.

> Real estate assets differ so greatly – how many models of f(*) do we need to have to account for all the differences.
>
> - y is the value output – the forever elusive 'real value'.
> - x is all the data input of the underlying asset.
> - The function f(*) is ultimately the valuation.
>
> (Xiong 2024)

AI hallucinates and, to avoid appearing ignorant, fabricates information AI hallucination is where a large language model '*perceives patterns or objects that are non-existent or imperceptible to human observers, creating outputs that are nonsensical or altogether inaccurate*' (IBM 2023).

Despite the success of Zillow in moving property attribute and value data online, the company suffered a significant setback when it decided to rely on an automatically generated value for making offers to purchase residential properties in the USA as part of its iBuyer programmes. Under the programme, Zillow purchased homes that were listed for sale with the intention that Zillow would then lightly renovate (if required) and then sold (known as flipping). In February 2022, Zillow reported an $881m loss for 2021 due to purchasing homes for prices higher than it believed it could then sell them for.

When commenting on the shortcomings of the Zillow model, Mike DelPrete (real estate technology strategist and scholar-in-residence at the University of Colorado in Boulder) said:

> you can have a real estate agent look at a house and in one second pick out one critical factor of the valuation that just doesn't exist as ones and zeroes in any database.

In undertaking a valuation, the inspection to identify critical factors would be undertaken by an appraiser (a valuer in Australia).

The quote from Xiong (2024), who is an advocate for AI, that '*AI hallucinates and, to seem not dumb, invents things*', is a critical shortcoming.

AI hallucination is a recognised issue and is an overwhelming reason that AI can only be a tool that is used under the total control of the Primary Valuer exercising professional scepticism.

The overwhelming shortcoming of AI is the limited nature or hygiene of a database. A valuer understands when sufficient relevant information has been compiled to form an opinion of value that is demonstrably objective, unbiased, ethical and competent.

Contrary viewpoints

Richard and David Susskind provide a viewpoint that is not accepted by the authors.

In 'The Future of the Professions', Richard and David Susskind present the case for the adoption of AI over the longer term (perhaps measured in decades). The immediate impact may be overestimated, but the impact of this technology over time may be underestimated. In a more recent article, they state (Susskind and Susskind 2023):

> There are of course risks with the latest AI. A recent technical paper on GPT4 acknowledges the systems can 'amplify biases and perpetuate stereotypes'. They can 'hallucinate'. They can also be plain wrong and raise the spectre of technological unemployment. Hence the frenzy of ethical and regulatory debate. At some stage, though, as its performance improves, and the benefits become unarguable, the threats and shortcomings will frequently be outweighed by the improved access AI provides.
>
> .... few professionals accept AI will take on their most complex work. They continue to imagine AI systems will be confined to their 'routine' activities, the straightforward, repetitive parts of their jobs — document review, administrative tasks, everyday grunt work. But for complex activities, many professionals argue that people will surely always want the personal attention of experts.

There is no doubt that AI will continue to advance, and the inputs will develop into more reliable outputs, hopefully removing hallucinations. AI, however, cannot inspect the property, the subject of the valuation. Further, a critical part of the valuation process is to obtain, compile, and appropriately verify sufficient relevant facts and data to ensure the valuer can conclude all reasoning and analysis to allow that valuer to reach an opinion of value, which is to inspect not only the subject property but also the evidence. Inspections reveal actual information at the date of inspection or the relevant date for the comparable evidence, whereas it is impossible for AI to know the specific detail at the relevant date.

Conclusion

AI will continue to be a very useful tool by creating cost/efficiency and time-saving in the valuation process, but will not replace the criterion that the value of the tangible asset is solely the unbiased opinion of the Primary Valuer.

Commercial pressures seek to hasten the adoption of AI technologies that can deliver faster and cheaper results. But the adoption of what are essentially proprietary 'black box' solutions is introducing opacity into a profession that is regulated and trusted (to the public advantage) by transparency. The lack of an ethical framework will undermine trust.

Scepticism is needed by those who rely on valuation services not to be fooled by the persuasive marketing of products that mimic a valuation but are created by Gen AI and fall short of the professional and regulatory standards that must be met by a professional valuer undertaking a valuation. Vendors should remain '*…vigilant against the dangers of over-reliance (driven by some very convincing sales pitches from vendors).*' (Bukowski-Kruszyna 2024)

As valuation work is often charged on a fixed fee basis, the improved profitability from possible efficiency savings provides a compelling case to adopt technology. However, artificial intelligence is a tool to be used for the valuer's benefit – it must not be implemented in such a way as to lead valuers into violating professional standards and ethics requirements, no matter how compelling the commercial case.

We need to learn from previous mistakes where technology has been unquestioningly accepted:

- Zillow – over US$800m in losses and an estimated 2,000 lost jobs due to reliance on an automated value and property purchase process.
- Robodebt – Commonwealth Ombudsman, Senate committee enquiries, and a Royal Commission resulted in 470,000 wrongly issued debts being refunded in full.
- Horizon – Royal Mail relied on software that resulted in more than 900 sub-postmasters being incorrectly prosecuted for stealing.

The introduction of AI to improve operational efficiency in the valuation or desktop process should be embraced by the professional valuer and valuation firms. The qualification, however, is that AI should not, in any way, influence the opinion of value by the Primary Valuer.

Selection of data using healthy scepticism meets the objective requirements of a valuer, whereas AI is subjective and limited to the quality and hygiene of the database, and, without knowledge that the content in the database is sufficient to form an unbiased opinion, the result may be erroneous. Automated processes do not apply professional scepticism to reject inaccurate or incomplete information held in a database.

In machine learning, the focus is on empirical performance, and there is less concern with model assumptions (Dunnigan 2024).

Whether in the longer term, AI can replicate the unbiased opinion of a valuer who has inspected the subject property in addition to the comparable evidence is yet to be seen. This is unlikely to happen in the medium term.

Bibliography

Adrian Wohlfarth (26 March 2024). Generative AI vs Traditional AI: Understanding the Difference https://www.ourcrowd.com/learn/generative-ai-vs-traditional-ai

Assessors and Registered Valuers Foundation (AaRVF), (n.d.). V20 Summit and Conference – Declaration Unifying VPOs for a Sustainable Future https://www.valuation20.org/v20india

Australian Property Institute. (2024, November 28). *Valuation process: Standards series No 3*. Retrieved October 10, 2025, from https://www.api.org.au/professional-development/knowledge-hub/valuation-process-is

Australian Property Institute Technical Working Group a subcommittee of the Standards Steering Committee (2023). (Kay, Hecek, Crawford, Preston, Brogan, Brandon) Response to IVSC Perspectives Paper Automated Valuation Models and Residential Valuations. https://www.api.org.au/wp-content/uploads/2023/05/API-response-to-IVSC-Perspectives-Paper-AVMs-17032023.pdf

Bell, The Hon AS Chief Justice of New South Wales (21 November 2024a). Supreme Court Practice Note SC Gen 23 – Use of Generative Artificial intelligence (Gen AI).

Bell, The Hon AS Chief Justice of New South Wales (4 December 2024b). 'Not a Substitute': Chief Justice Responds to Criticisms of GenAI Ban.

Bukowski-Kruszyna, Konrad (27 May 2024). Institute of Chartered Accountants for England and Wales 'Over-Reliance on Automation: A Cautionary Tale from Plato' https://www.icaew.com/insights/viewpoints-on-the-news/2024/may-2024/overreliance-on-automation-a-cautionary-tale-from-plato?utm_campaign

Chien, Colleen; Kim, Miriam, *Generative AI and Legal Aid: Results from a Field Study and 100 Use Cases to Bridge the Access to Justice Gap*' (Research Paper forthcoming, Berkeley School of Law, University of California, 11 April 2024).

Assessors and Registered Valuers Foundation. (2023). *India's economic perspectives and progress: Valuation 20 Conference Proceedings 2023*. AaRVF Digital Journal. Retrieved October 11, 2025, from https://journals.aarvf.org/v20publication-papers

'Digital Planning Programme Funding 2023'. https://www.localdigital.gov.uk/digital-planning/funding/digital-planning-programme-funding-2023/

Dunnigan, Gerri (17 December 2024). *Machine Learning vs Statistics: What's the Best Approach*. University of North Dakota. https://und.edu/blog/machine-learning-vs-statistics.html#:~:text=In%20statistics%2C%20the%20methodology%20emphasizes,less%20concern%20with%20model%20assumptions

FIG Commission 9 (n.d.). Valuation and the Management of Real Estate. https://www.fig.net/organisation/comm/9/

Forlee, Ron (5 July 2022). *Real Estate Development Strategy for Investors* https://books.google.com/books?id=GmF6EAAAQBAJ&dq=proptech&pg=PA217. John Wiley & Sons. p. 219. ISBN 978-1-119-88732-4

International Valuation Standards Council (2024). International Valuation Standards Effective 31 January 2025: https://saicawebprstorage.blob.core.windows.net/uploads/resources/IVS-effective-31-January-2025.pdf

IBM. (2023). What are AI Hallucinations? IBM. Retrieved October 11, 2025, from https://www.ibm.com/think/insights/ai-hallucinations-pose-risk-cybersecurity

Jones, Colin A.; Trevillion, Edward (18 August 2022). *Real Estate Investment: Theory and Practice* . Springer Nature. pp. 288–290. ISBN 978-3-031-00968-6. https://books.google.com/books?id=6SKDEAAAQBAJ&dq=proptech&pg=PA286

Liaw, K. Thomas (14 June 2021). *The Routledge Handbook of FinTech*. Routledge. ISBN 978-1-000-37573-2. https://books.google.com/books?id=ke0rEAAAQBAJ&dq=proptech&pg=PT428 ISBN 978-1-000-37573-2

Metz, Rachel (9 November 2021). Zillow's Home-buying Debacle Shows How Hard It Is to Use AI to Value Real Estate. https://edition.cnn.com/2021/11/09/tech/zillow-ibuying-home-zestimate/index.html

'PropTech Engagement Fund Round 1' (n.d.). https://www.localdigital.gov.uk/fund/proptech-engagement-fund/

Queensland Law Society – AI Companion Guide (25 October 2024).

Reed, Richard (30 March 2021). *Property Development* (https://books.google.com/books?id=z3QhEAAAQBAJ&dq=proptech&pg=PT330). Routledge. ISBN 978-1-000-35994-7

Rishi, Sanjay; Breslau, Benjamin; Miscovich, Peter (29 October 2021). *The Workplace You Need Now: Shaping Spaces for the Future of Work* (https://books.google.com/books?id=2m9LEAAAQBAJ&dq=proptech&pg=PA146 John Wiley & Sons. p. 146. ISBN 978-1-119-81512-9

Schopf, S. Surrogate's Court Saratoga County. In the Matter of the Accounting by Susan F. Weber as the Trustee of the Michael S. Weber Trust for the benefit of Owen K. Weber under Article SEVENTH of the Last Will and Testament of Michael S. Weber, Decedent. Decided on 10 October 2024.

Soules, Matthew (4 May 2021). *Icebergs, Zombies, and the Ultra-Thin: Architecture and Capitalism in the 21st Century* . Chronicle Books. p. 205. ISBN 978-1-64896-029-1. https://books.google.com/books?id=vHQOEAAAQBAJ&dq=proptech&pg=PA205/

Susskind, Richard; Susskind, Daniel (June 2023). Generative AI Will Upend the Professions. https://www.ft.com/content/96a1877f-0bbb-48c7-be8f-4fed437810e8?accessToken=zwAF_mRYR6-YkdOWoYd_C7tIx9O-j0_tQ3gQ6A.MEQCIELVeuT1USAktchHVGCEvXzg_7fO1KuwUg6xnzGr6iWAAiAUyO4KERQsSUl13RPMcFv86MxTIzfSIJT4elpO_mYGCw&sharetype=gift&token=e91fff15-d799-4ace-a830-8d3d8ab09059

Swift, Jonathan (1725). Gulliver's Travels into Several Remote Nations of the World (Project Guttenberg) https://www.gutenberg.org/files/829/829-h/829-h.htm

Tiwari, Piyush; Miao, Julie (8 March 2022). *A Research Agenda for Real Estate* . Edward Elgar Publishing. pp. 25–30. ISBN 978-1-83910-393-3. https://books.google.com/books?id=Yq1jEAAAQBAJ&dq=proptech&pg=PA25

Voices, Valley (2019). Modernizing Real Estate: The Property Tech Opportunity. *Forbes*. Retrieved 2020-07-23. https://www.forbes.com/sites/valleyvoices/2019/02/22/the-proptech-opportunity/

Wikipedia (4 December 2024). Property technology https://en.wikipedia.org/wiki/Property_technology2/

Wohlfarth, A. (2024, March 26). Generative AI vs. traditional AI: Understanding the differences. *OurCrowd.* Retrieved October 10, 2025, from https://www.ourcrowd.com/learn/generative-ai-vs-traditional-ai

Xiong, L. Q. (2024, October 24). Ethics in the age of AI: Impact on valuation practices [Panel debate]. The CPD Foundation Harris Debate, Royal Institution of Chartered Surveyors, London, United Kingdom. Retrieved October 11, 2025, from https://jonathandavidharris.info/the-rics/

Chapter 3

Artificial intelligence and machine learning in property and real asset valuation

Policies, Practices and Emerging Trends

Isara Khanjanasthiti, Samuel Gardner and Armin Taklif

Introduction

Valuation is the process of determining a property's or real asset's current worth, undertaken by an appraiser who evaluates various factors to estimate its value. Systematically, valuation has substantially evolved from its origin, during which subjective assessments and rudimentary estimations, which are prone to inaccuracy and unreliability, were common. Nowadays, a valuer's considerations are comprehensive and typically include the property's or asset's location, size, condition and recent sales of similar properties nearby. The aim of conducting a valuation is to provide an accurate and fair assessment of 'value', which can be used for various purposes such as buying, selling, taxation and aiding in investment decisions.

Artificial Intelligence (AI) has recently received significant publicity, largely due to the public release of generative AI models. These models, which continue to evolve, use deep learning techniques to analyse patterns within vast datasets and generate original content. They rely on neural networks trained on diverse sources to produce contextually relevant outputs based on user inputs. They are also capable of imitating human creativity and thought processes, offering significant possibilities for innovation and efficiency in various fields. Among these fields, the built environment industry has been progressively embracing digital technologies, including AI and ML, in recent years.

AI and machine learning (ML) technologies, such as automated valuation models (AVMs) and predictive analytics, have demonstrated significant potential in automating valuation processes, generating real-time insights and enhancing decision-making capabilities (Cajias, 2021; Ja'afar et al., 2021; Su et al., 2021). As such, this chapter explores the transformative potential of AI and ML on property and asset valuation, providing a comprehensive analysis of how these technologies can enhance accuracy, efficiency and sustainability in the property sector and highlighting the importance of enabling regulatory environments. The chapter argues that the integration of AI and ML into property and asset valuation can considerably enhance accuracy and efficiency.

DOI: 10.1201/9781003539254-4

However, robust policy frameworks and collaboration are required to ensure ethical, transparent and sustainable practices.

Three objectives guide this chapter: (1) to review the development and application of AI and ML in valuation, (2) to highlight the alignment of AI- and ML-driven valuation tools with the United Nations Sustainable Development Goals (SDGs) and (3) to discuss the importance of policy and regulatory environments in facilitating AI- and ML-driven valuation. To achieve these objectives, the chapter begins with a review of the current literature on the application of AI and ML in property valuation, examining global best practices that have emerged in recent years. The literature review involved a targeted search and critical review of relevant peer-reviewed articles, industry reports, government publications, international standards and regulatory frameworks.

Following the literature review, the chapter explores policy frameworks and strategic objectives that support the adoption of these technologies in property and asset valuation. The discussion draws on examples from regions where these technologies have been successfully integrated into regulatory environments, emphasising the importance of government support and industry collaboration (KPMG, 2023). The alignment of AI and ML with SDGs 9 (Industry, Innovation and Infrastructure) and 11 (Sustainable Cities and Communities) is highlighted, showcasing their potential to contribute to sustainable urban development (United Nations, 2015).

The chapter then examines two case studies where AI and ML have been successfully implemented in property and asset valuation. These case studies provide real-world examples of the benefits these technologies offer, including improved accuracy, cost and time savings (Baum et al., 2021). By examining these implementations, the chapter identifies key challenges and opportunities, such as the need for data integration, technological infrastructure and the development of skilled professionals to interpret AI-driven insights. Finally, the chapter concludes with recommendations for stakeholders in the property sector. These recommendations focus on fostering collaboration between academia, industry and government to fully realise the potential of AI and ML. A proposed policy framework can guide the integration of these technologies into valuation practices, ensuring their effective use in driving innovation, sustainability and resilience in the built environment (Rathore et al., 2021).

The evolution of valuation practices

Overview of artificial intelligence and machine learning

According to IBM (2023), AI refers to *'machines that mimic human intelligence and human cognitive functions like problem-solving and learning [by using] predictions and automation to optimise and solve complex tasks that humans have historically done, such as facial and speech recognition, decision-making and translation'*. These AI systems leverage vast amounts of data and advanced algorithms to continuously improve their performance and accuracy over time. There are different types of AI systems (e.g., natural language processing, computer vision and robotics), each serving a different purpose and relying on distinct types of data. In recent years, significant progress has been made in the capabilities of AI, enabling it to consistently outperform humans on several benchmarks and driving adoption by both industry sectors and the general public (Stanford Institute for Human-Centered Artificial Intelligence, 2024).

ML is a subset of AI that automates analytical model building. By learning from large data sets, ML systems can identify patterns and make decisions with limited human intervention. Three major types of ML systems include 1) supervised learning (a model being trained on labelled data), 2) unsupervised learning (a model identifying patterns in unlabelled data) and 3) reinforcement learning (a model learning by interacting with its surroundings and receiving feedback) (IBM, 2023).

Nowadays, AI and ML are pivotal technologies across various industries, underpinning their innovation and efficiency. They have also received substantial research and development, with the Stanford Institute for Human-Centered Artificial Intelligence (2024) highlighting a 31-time increase in AI patents between 2010 and 2024, an introduction of 88 notable ML models in 2023 and rapidly growing AI research projects and publications.

How artificial intelligence and machine learning are transforming valuation

In the context of property valuation, AI and ML can be described as 'machine learning designed to predict an outcome or provide an estimate', with an example being the 'most probable sale price' (Gloudemans & Sanderson, 2021, p. 90). This section explores the different transformations that have been seen since the application of AI and ML to mass appraisals and property valuations. It also identifies key trends in the uptake of AI developments and software across the public and private sectors.

The International Valuation Standards Council (IVSC, 2024, p. 9) defines 'market value' as:

> The estimated amount for which an asset or liability should exchange on the valuation date between a willing buyer and a willing seller in an arm's length transaction after proper marketing where the parties had each acted knowledgably, prudently and without compulsion.

This definition highlights three main requirements for a market transaction to occur: 1) a 'willing seller' and 2) a 'willing buyer', both of whom have acted 3) 'knowledgeably, prudently and without compulsion' in agreeing to a sale price in an 'arm's length transaction'. According to the Royal Institution of

Chartered Surveyors (RICS, 2019), property valuers are required to analyse recent sales transactions and provide educated inputs (sales evidence) to gauge the market value of the subject property.

The primary distinction between traditional valuations and AI- and ML-driven valuations is that while the former depends on the valuer's critical analysis of available evidence to construct a valuation of the subject property, the latter heavily relies on refined, existing databases of relevant information. AI and ML have experienced difficulties aligning values with traditional property valuations due to the heavy reliance on individual valuers to provide educated inputs, opinions of value and professional judgement for the subject property. Early AI applications in valuation could not recreate professional judgement and other personal opinions when undertaking a market valuation on a subject property. Nevertheless, modern AI tools such as AVMs have been able to recreate the opinions of valuers through the pairing of adaptive learning models and the Hedonic Price Model (HPM) to recreate certain decisions by a property valuer (Tchuente, 2024). The RICS (2021, p. 5) provides the following definition of AVMs while also highlighting the challenge of lacking a universally agreed-upon definition:

> Automated Valuation Models use one or more mathematical techniques to provide an estimate of value of a specified property at a specified date, accompanied by a measure of confidence in the accuracy of the result, without human intervention post-initiation.

By combining mathematical and statistical modelling techniques with comprehensive databases of existing properties and transactions to calculate real estate values, AVMs are used to estimate the market value of real properties. AVMs are widely used in mortgage lending, statutory tax assessments and online real estate sales platforms for their efficiency and consistency compared to human appraisers (Glumac & Des Rosiers, 2021). AVMs are now leveraging AI to refine data inputs for adaptive ML algorithms. This data drives accurate outputs and assists in troubleshooting errors within modern AVMs. AI and ML technologies are also being used to develop datasets that incorporate macroeconomic and demographic data to refine hedonic price model outputs (Brown & Engler, 2023).

Advantages and disadvantages of AI- and ML-driven valuation

Below are the key advantages of valuation tools utilising AI and ML technologies (Baum et al., 2021):

1 **Efficiency**: In comparison to traditional valuations, AI- and ML-driven valuation tools such as AVMs provide a quick and effortless method of completing large volumes of valuations in a shorter period for a low to moderate

cost. Major developments have been made in the ML models and the implementation of HPM and artificial neural network models (ANN). These methods are concentrated on the direct comparison approach, as it is an intuitive, non-scientific process performed by an experienced valuer.

2 **Reduced susceptibility to human bias**: The direct comparison approach considers key characteristics such as the location and size of the land, the level of improvements such as the number of bedrooms, bathrooms and car spaces and the location and its proximity to key social infrastructure. AI- and ML-driven valuations can reduce human bias and external influences to provide a data-driven, less biased valuation. The broader adoption of AVMs has the potential to limit fraudulent valuations that are subjective to external client influence. However, while the reduction in human intervention in the valuation process of AVMs can make the value estimates more precise, these estimations are still subject to 'bias'. This is due to factors such as algorithm design and dataset quality (Topraklı, 2024).
3 **Multi-sectoral application**: AI- and ML-driven valuations will provide an opportunity for both public and private implementations to undertake large-scale valuations in a short period of time. Public applications could include the use of AVMs for land tax valuations for government departments internationally. Meanwhile, private applications may focus on underwriting mortgages and insurance with specific advantages for large-scale portfolio valuations, with future development of AVMs to include risk and liquidity analysis for forecasting downturns within the market.

However, the application of these technologies in property valuation is not without flaws, with the following three key disadvantages:

1 **'Black box' nature**: As most AI and ML-driven valuation tools have been developed in the private sector, there are issues surrounding intellectual property and transparency of algorithmic calculations used in such tools. AVMs can be broadly broken down into three key steps. The first step involves embedding inputs, including datasets and variables, into the models. The second step is the process that varies between tools and models and is based on algorithms protected as intellectual property. Lastly, the output is widely shared with clients and users to identify the performance of the AVMs. Key issues arise around the transparency of data inputs and outputs, the employed algorithms and the difficulty in model explanation and interpretation. These factors have hindered trust and broader adoption of AVMs in valuation (Cheung & Yiu, 2024; Topraklı, 2024).
2 **Susceptibility to dataset bias**: Quality datasets are crucial for developing AI- and ML-driven valuation tools, given their reliance on accurate and dependable data to produce precise valuation outputs (Pagano et al., 2023). However, dataset bias is found when several anonymities are present, such as a lack of available transaction data, small sample sizes and sales data not representing current market parameters (i.e., transaction values do not

accurately reflect market values). This becomes problematic when markets experience sudden movements in prices or a surge in supply or demand.

3 **Lack of 'human' perceptions**: AI- and ML-driven valuation tools have a major disadvantage in understanding external factors and the influence of human behaviours in the property market, which can often be unpredictable. An example of this disadvantage is calculating valuations in markets with limited comparable assets in niche geographical areas where an asset is either above or under the market's expectations (Jafary et al., 2022). Valuation tools such as AVMs have limited ability to replicate the 'human touch' needed for valuations that require diverse inputs and critical thinking to confirm an opinion of value.

Alignment with Sustainable Development Goals

AI- and ML-driven valuation tools can potentially assist decision-makers for cities and towns in achieving two of the 17 SDGs, namely SDG 9 and SDG 11.

Goal 9: Industry, Innovation and Infrastructure

SDG 9 (Industry, Innovation and Infrastructure) aims to 'build resilient infrastructure, promote inclusive and sustainable industrialisation and foster innovation' and is underpinned by the following eight targets (United Nations, 2015, pp. 22–23):

9.1. Develop quality, reliable, sustainable and resilient infrastructure, including regional and transborder infrastructure, to support economic development and human well-being, with a focus on affordable and equitable access for all;

9.2. Promote inclusive and sustainable industrialisation and, by 2030, significantly raise industry's share of employment and gross domestic product, in line with national circumstances, and double its share in least developed countries;

9.3. Increase the access of small and medium-sized industrial and other enterprises, in particular in developing countries, to financial services, including affordable credit, and their integration into value chains and markets;

9.4. By 2030, upgrade infrastructure and retrofit industries to make them sustainable, with increased resource use efficiency and greater adoption of clean and environmentally sound technologies and industrial processes, with all countries taking action in accordance with their respective capabilities;

9.5. Enhance scientific research, upgrade the technological capabilities of industrial sectors in all countries, in particular developing countries, including, by 2030, encouraging innovation and substantially increasing the number of research and development workers per 1 million people and public and private research and development spending;

9.a. Facilitate sustainable and resilient infrastructure development in developing countries through enhanced financial, technological and technical support to African countries, least developed countries, landlocked developing countries and small island developing States;
9.b. Support domestic technology development, research and innovation in developing countries, including by ensuring a conducive policy environment for, *inter alia*, industrial diversification and value addition to commodities; and
9.c. Significantly increase access to information and communications technology and strive to provide universal and affordable access to the Internet in least developed countries by 2020.

AI and ML technologies can directly contribute to SDG 9 by fostering resilient infrastructure, inclusive industrialisation and innovation in the following ways:

- **Resilient Infrastructure**: AI-driven tools, such as predictive analytics, enhance resilience by enabling stakeholders to anticipate and mitigate economic and environmental risks. These tools also optimise resource allocation, promoting long-term sustainability (Comunale & Manera, 2024).
- **Inclusive Industrialisation**: AI expands access to advanced property valuation tools for under-resourced regions, democratising access to technology and enabling equitable industrial growth (United Nations, 2024).
- **Innovation**: By streamlining traditional valuation practices, AI fosters innovation across the property sector. AVMs can enhance accuracy and efficiency, enabling data-driven decision-making (OECD, 2024a).

The United Nations (2024) emphasises the transformative potential of AI in achieving SDGs, highlighting how AI can modernise industries, foster inclusive growth and support sustainable infrastructure development, particularly in developing nations. Through AI and ML applications, the property sector can integrate sustainability into valuation practices, advancing the industry towards both economic and environmental resilience.

Goal 11: Sustainable Cities and Communities

SDG 11 (Sustainable Cities and Communities) aims to 'make cities and human settlements inclusive, safe, resilient and sustainable' by 2030 and is underpinned by the following ten targets (United Nations, 2015, p. 24):

11.1. By 2030, ensure access for all to adequate, safe and affordable housing and basic services and upgrade slums;
11.2. By 2030, provide access to safe, affordable, accessible and sustainable transport systems for all, improving road safety, notably by expanding public transport, with special attention to the needs of those in vulnerable situations, women, children, persons with disabilities and older persons;
11.3. By 2030, enhance inclusive and sustainable urbanisation and capacity for participatory, integrated and sustainable human settlement planning and management in all countries;
11.4. Strengthen efforts to protect and safeguard the world's cultural and natural heritage;
11.5. By 2030, significantly reduce the number of deaths and the number of people affected and substantially decrease the direct economic losses relative to global gross domestic product caused by disasters, including water-related disasters, with a focus on protecting the poor and people in vulnerable situations;
11.6. By 2030, reduce the adverse per capita environmental impact of cities, including by paying special attention to air quality and municipal and other waste management;
11.7. By 2030, provide universal access to safe, inclusive and accessible, green and public spaces, in particular for women and children, older persons and persons with disabilities;

11.a Support positive economic, social and environmental links between urban, peri-urban and rural areas by strengthening national and regional development planning;
11.b By 2020, substantially increase the number of cities and human settlements adopting and implementing integrated policies and plans towards inclusion, resource efficiency, mitigation and adaptation to climate change, resilience to disasters, and develop and implement, in line with the Sendai Framework for Disaster Risk Reduction 2015–2030, holistic disaster risk management at all levels; and
11.c Support least developed countries, including through financial and technical assistance, in building sustainable and resilient buildings utilising local materials.

Government agencies can employ AI-driven property valuation tools to acquire more accurate and timely land and property valuations. More accurate and timely property valuations are essential for countries where land and property are major asset classes. For instance, Australia, as a country in which land and property are the most predominant asset classes (valued at $6.8 trillion), collects more than $50 billion annually in property tax (University of New South Wales, n.d.).

More accurate and timely land and property valuations can lead to increased government tax revenue and levies from these assets, which can be spent on funding infrastructure and services for cities (Burton, 2023; Jafary et al., 2024). Funding the appropriate infrastructure and services for cities can directly assist city administrators in achieving the following targets of SDG 11 (United Nations, 2015, p. 24):

- 11.1 ('adequate, safe and affordable housing and basic services and upgrade slums');
- 11.2 ('safe, affordable, accessible and sustainable transport systems');
- 11.6 ('air quality and municipal and other waste management'); and
- 11.7 ('safe, inclusive and accessible, green and public spaces').

Access to accurate land prices enabled by AI property valuation tools can assist in government agencies' urban development and planning efforts, thus enabling them to broadly address SDG 11. Importantly, policymakers, urban planners and other stakeholders can also use such data to spatially identify suburbs requiring more affordable housing and introduce housing or land use initiatives accordingly (Jafary et al., 2024). This approach is consistent with target 11.1 of SDG 11.

Valuation practices, traditionally reliant on human input, are poised for a revolution driven by AI and ML technologies, bringing various potential benefits. However, as illustrated above, these technologies are subject to potential shortcomings, highlighting the important role of conducive policy and regulatory environments, which are discussed in the next section of the chapter.

The importance of policy and regulatory environments

Existing regulations around valuation

The Australian Property Institute's (API) *Rules of Professional Conduct* outline the processes for undertaking valuations within Australia. This document runs parallel to the requirements of the *International Valuation Standards* (IVS) published by the International Valuation Standards Council (IVSC), given that the IVS have been 'adopted in full' by the API (API, 2024, p. 10). In line with the API's standards, physical inspections are required when undertaking a valuation of tangible assets such as real property and plant, machinery and equipment. Currently, the following three forms of physical inspections are approved by the API (2024):

- Comprehensive physical inspection;
- Limited (or partial) on-site inspection; and
- Limited off-site (kerbside) inspection.

For valuation on assets that are not comprehensively inspected (e.g., 'desktop' valuations), the valuer is required to have received written instructions from the client (API, 2024). The API (2024, p. 3) defines 'desktop' as

> An evidence-based desktop process for assessing the value of a tangible asset as at a specified date but does not include a physical inspection (of any kind) of the asset by the Valuer.

Due to their nature of not involving physical inspections by valuers, desktop valuations of tangible assets strongly parallel AVMs. It should be noted, however, that this form of valuation is not considered compliant under the IVS. Therefore, it can only be used to provide an *indicative* value. As such, for this type of valuation, the API has emphasised that the valuer is required to have sufficient knowledge to ascertain and verify sufficient evidence to complete the valuation (API, 2024). Overall, the sourcing of sufficient and reliable evidence is paramount to providing a precise, indicative valuation of the subject asset.

Internationally, the RICS published a road map in mid-2021 to address AVMs' development, adoption and understanding. The RICS provided a white paper addressing current regulations and the need for due diligence and cooperation across the property profession (Knight, 2022). This paper indicates that developed markets in North America, the United Kingdom and Australia were the first adopters of hybrid AVMs. This form of AVM employs a defined database to undertake analysis through a hedonic model to provide an estimated value for an asset. The hybrid model requires a valuer to undertake an analysis of results and to perform a critical review of the outlying assessments of the model (Baum et al., 2021). This practice has become popular due to the focus of AVMs on the residential markets, with recent developments being made in the implementation of AVMs for the mass appraisal of land for property taxation in the public sector (Knight, 2022).

The rapid development and application of AI and ML in property and asset valuation hold substantial potential to revolutionise traditional valuation methods. However, the success and sustainability of these technologies depend heavily on the establishment of robust policy and regulatory environments. Such frameworks are essential for ensuring the ethical, secure and effective integration of AI and ML into industry practices. Without appropriate regulations and supportive policies, these technologies risk misuse, data privacy breaches and erosion of public trust, all of which could impede widespread adoption (Pantanowitz et al., 2024; Yanamala & Suryadevara, 2024).

Key attributes of conducive policy and regulatory environments

A conducive policy and regulatory environment for AI- and ML-driven property valuation must encompass the following key attributes.

Proactive professional bodies

Professional bodies play a critical role in promoting the adoption of AI and ML property valuation methods amongst practising valuers (Abidoye et al., 2021). Despite the potential of AI in property valuation, in practice, valuers have shown hesitation in adopting AI-driven property valuation methods. Based on a survey of valuers practising in Australia, Abidoye et al. (2021) propose that there are mixed opinions on adopting AI property valuation methods amongst Australian valuers, with some viewing it as transformative and others more sceptical. They highlight the role of valuers' professional bodies in promoting AI adoption through key mechanisms such as workshops and continuing professional development (CPD) activities.

Ethical standards and guidelines

Ethical considerations are essential to deploying AI and ML technologies responsibly. Policies must include explicit ethical standards to prevent these technologies from perpetuating biases or discriminating against specific groups. For instance, if left unchecked, AI algorithms can reinforce societal inequities by amplifying biased datasets. A robust policy framework should mandate transparency in algorithm development and data usage, requiring developers of AI- and ML-driven valuation models to explain how their models function and make decisions (Floridi et al., 2018; Pantanowitz et al., 2024).

Additionally, policies and guidelines mandating the disclosure of algorithms and training data used in developing AI property valuation tools are essential for ensuring transparency to users (Ota & Uto, 2024). Ota and Uto (2024) explored five AI property valuation services and evaluated 4295 valuations in 859 condominium units in Tokyo from these tools and discovered an average variation of 10.6% in the valuations. This variation is significantly higher than that seen in traditional property valuations, emphasising the need for transparency in algorithms and training data to allow for a more thorough examination of the calculations used by these tools.

Data privacy and protection regulations

Given the data-intensive nature of AI and ML, safeguarding privacy is critical. Regulatory frameworks must enforce stringent data protection standards that align with global norms, such as the *General Data Protection Regulation* (GDPR) in the European Union. These measures should ensure that personal and property-related data are responsibly collected, stored, and processed, preserving individuals' rights while upholding the integrity of valuation models (European Commission, n.d.; Yanamala & Suryadevara, 2024).

Standardisation and interoperability

The establishment of industry-wide standards is vital for enabling interoperability across AI and ML systems. Policies should promote the adoption of standardised data formats and protocols to facilitate seamless integration between different technologies. According to the International Organization for Standardization (ISO, 2022), such standardisation encourages collaboration, improves efficiency and promotes the exchange of best practices across sectors.

Support for innovation and research

A balanced policy framework must encourage innovation by providing incentives and funding for research and development in AI and ML applications specific to property and asset valuation. By fostering partnerships between academia, industry and government, these policies can stimulate technological progress while ensuring alignment with societal goals (Aldoseri et al., 2024; Pantanowitz et al., 2024).

Flexibility and adaptability

Given the rapid evolution of digital technologies, overly rigid policies can stifle innovation. Regulatory frameworks must include mechanisms for periodic review and updates to address emerging challenges and opportunities, such as advancements in AI algorithms or data management techniques. For example, incorporating regular assessments of new AI applications can address unforeseen ethical concerns while maintaining a focus on innovation (Lescrauwaet et al., 2022; OECD, 2024b).

Workforce development and training

The integration of AI and ML technologies requires not only adaptable frameworks but also a skilled workforce capable of navigating these changes. Policies should incentivise training programmes that equip professionals with the technical expertise and ethical awareness needed to leverage AI and ML effectively. Such programmes should emphasise cross-disciplinary collaboration and promote technical proficiency (Lescrauwaet et al., 2022; OECD, 2024b).

Collectively, these attributes form the foundation of a regulatory environment that fosters the ethical and effective use of AI and ML in property valuation. Understanding how these principles operate in practice requires examining real-world examples of successful policy frameworks, which are discussed next.

Examples of 'Best Practices' in policy and regulatory environments

This section highlights eight examples of 'best practices' in promoting policy and regulatory environments that are conducive to the implementation of AI- and ML-driven valuation tools to date.

European Union's Artificial Intelligence Act

The European Union's *Artificial Intelligence Act* exemplifies a balanced approach to innovation and regulation. It categorises AI systems by risk level, from minimal to high risk, and imposes corresponding regulatory requirements. 'High-risk' AI systems – including AVMs – must meet stringent criteria for data governance, transparency and human oversight. This framework ensures that AI use is ethical and transparent, mitigating risks while encouraging innovation (European Commission, 2021).

Singapore's AI Governance Framework

Singapore's draft *Model AI Governance Framework for Generative AI* prioritises accountability and transparency, requiring organisations to implement structures that oversee AI deployment. The framework promotes explainable AI (XAI), ensuring that AI-driven decisions, such as property valuations, are easily understood by stakeholders. This has fostered trust and accelerated the integration of AI in the real estate and finance sectors (Infocomm Media Development Authority, 2024).

United States' Federal Initiatives and State Policies

The United States employs a decentralised approach to AI regulation, with federal agencies and state governments driving sector-specific policies. For instance, the Federal Trade Commission enforces consumer protection in AI applications, while the *California Consumer Privacy Act* imposes stringent rules on data collection and opt-out rights. These layered efforts address ethical considerations while supporting technological progress ('Biometric Information Privacy Act', 2008; Kosinski, 2024).

Australia's Approach to AI Ethics and Standards

The Australian Government's (2024) *AI Ethics Principles* provides voluntary guidelines emphasising fairness, privacy and reliability. Amongst the eight principles outlined, 'reliability and safety' and 'transparency and explainability' are particularly relevant to AI- and ML-driven valuation tools due to their 'black box' issues. Adhering to these principles ensures that AI systems in property valuation maintain public trust while contributing positively to societal outcomes. Pilot programmes exploring the integration of AI in urban planning and property management further demonstrate Australia's proactive regulatory approach (New South Wales Government, 2024; Pettit, n.d.).

United Kingdom's Centre for Data Ethics and Innovation

The United Kingdom's Centre for Data Ethics and Innovation (CDEI) plays a pivotal role in shaping AI policy and ensuring that technological innovation aligns with public interest. By conducting research and providing recommendations on data ethics, the CDEI has influenced policies that promote ethical AI practices. This has directly supported advancements in property valuation while maintaining transparency and accountability (Farrow et al., 2022).

China's Integrated AI Regulation Strategy

China is emerging as a global leader in AI innovation, with a focus on dominance in the sector while maintaining strict ethical and security standards. The *New Generation Artificial Intelligence Development Plan* aims to integrate AI across multiple industries and establish regulations to ensure responsible development. More recently, the *Interim Measures for the Management of Generative AI Services* document outlines clear guidelines for data protection, transparency and accountability. One of the key requirements states that 'effective measures must be used to increase the transparency, accuracy, and reliability of generative AI services' (Ferguson et al., 2023). Given the 'black box' nature of AI- and ML-driven valuation tools, this requirement is vital in enhancing their transparency, which is essential for building trust and encouraging wider adoption within the real estate sector.

Canada's Proactive AI Governance

Canada employs a balanced approach to fostering innovation while upholding ethical standards. The *Pan-Canadian Artificial Intelligence Strategy* promotes responsible AI adoption, while the *Artificial Intelligence and Data Act* ensures transparency, accountability and data protection. Together, these initiatives position Canada as a leader in AI-driven innovation (Government of Canada, 2023).

Policy Integration with International Standards

International standards are crucial in ensuring global interoperability and addressing gaps in national AI policies. Frameworks such as the OECD AI Principles (OECD, 2024a) and ISO/IEC 22989 (ISO, 2022) establish a foundation for transparency, accountability and inclusivity in AI governance. These standards enable seamless integration of AI systems across borders, facilitating collaboration and reducing barriers to innovation. While regional regulations such as the GDPR prioritise data privacy within specific jurisdictions, international standards provide guidelines for addressing broader challenges, such as cross-border data sharing and algorithmic accountability. This global consistency is especially critical in property valuation, where AI technologies often operate across diverse legal and cultural contexts. Aligning national policies with international standards also supports global initiatives such as the United Nations SDGs (United Nations, 2024). By prioritising fairness and inclusivity, nations can ensure that AI technologies contribute to equitable and sustainable development worldwide.

The diverse policy case studies above illustrate that jurisdictions have adopted key shared principles to guide the responsible governance of AI systems. Transparency is a core requirement across regions, enabling stakeholders to understand and trust AI- and ML-driven property valuation tools. Risk categorisation informs appropriate regulatory responses based on the level of risk posed by AI systems. Accountability measures are also critical to supporting the ethical use and outcomes of these technologies. In addition, jurisdictions consistently emphasise data protection, reliability and the alignment of AI practices with broader societal values, such as fairness and inclusivity. International frameworks further promote consistency in AI governance across borders, supporting collaboration and interoperability. Together, these shared principles create a conducive policy environment that balances innovation with public trust, enabling the responsible integration of AI- and ML-driven property valuation tools.

Having examined the importance of conducive policy and regulatory environments in enabling the use of AI- and ML-driven valuation tools, the next part of the chapter outlines two case studies illustrating a successful implementation of these tools.

Case studies of successful implementation of AI- and ML-driven valuations

This section outlines and discusses two case studies of successful AI- and ML-driven valuations. Each case study begins with an overview, followed by a discussion of the policy and regulatory environments that enable the development and use of AI tools. Finally, key outcomes and lessons from the implementation of these tools are examined.

Landgate, Western Australia

Overview

Landgate is a commercial statutory authority established by the Western Australian Government in 2007 after the State Parliament passed the *Land Information Authority Act* in 2006. The authority was founded to implement new technologies and streamline operations and processes across the state's land titling, land valuation and spatial data divisions.

Enabling policy and regulatory environments

The *Land Information Authority Act 2006* provided an opportunity for the streamlining of land management professionals under one entity. In comparison to their eastern counterparts that manage land registry operations and statutory valuations in a fragmented manner through numerous state government departments, Landgate was able to establish a collection of interconnected professionals (Gainer, 2017).

Since its establishment, Landgate has undergone a transition away from the traditional paper-based land registration system to a fully digitalised system. This transformation has enabled investment in the rapid development of technology and critical database infrastructure (Gainer, 2017). This has been aided by the commercial powers permitted to the organisation under the statutory authority model.

The *Land Information Act 2006* is reviewed every five years to 'ensure Landgate's operations are effective and meet the needs of the State', with the last review completed in 2022 (Landgate, 2024b). A board, operating within a traditional commercial structure, manages ongoing operations and reports to the Minister of Lands. This structure allows for greater agility in investing in new software and managing commercial operations compared to a traditional government department. Consequently, Landgate has rapidly implemented new technologies such as AI and ML, providing valuable insights to the minister for policies affecting the organisation's ongoing operations and development (Landgate, 2024b).

Outcomes and lessons

Technology development is a key objective of Landgate, and a major project includes developing the *New Land Registry* (NLR) system. This platform significantly reduced the time required for land sale registration. Using a traditional system, the process took seven to ten days. With the new platform, paper-based transactions now take one and a half days, and electronic transactions are completed in only *23 seconds* (Gainer, 2017). Additionally, with 1,530,415 Certificates of Title in Landgate's register in the year ending 30 June 2024 (Landgate, 2024a), AI and ML have played a critical role in improving the agency's efficiency in its valuation process, with no limitation of location and time. One significant benefit of the platform's paperless transactions is its contribution to removing the need for more than 5,000,000 pieces of paper between 2016 and 2024 (Landgate, 2024a). This is essential in helping the Western Australian Government achieve its ambitious environmental sustainability goal of net zero for the 'emissions profile of Western Australia and governance framework' by 2050 (Western Australian Government, 2024, p. 4).

Due to its commercial nature, Landgate has been able to develop and implement valuation systems for the Valuer-General of Western Australia to undertake statutory valuations. Landgate has used computer-assisted mass appraisal software and AVMs to produce accurate and cost-effective statutory valuations. For the year ending 30 June 2024, Landgate valued over a million properties throughout Western Australia. This valuation provided essential data for local council rates, land tax and other state government charges and levies (Landgate, 2024a).

The case study of Landgate highlights the advantages commercial statutory authorities have in implementing new technology and streamlining government functions. Since its establishment, Landgate has been at the forefront of developing and investing in property-based software while fulfilling its statutory obligations to the Minister of Lands and the state government. This has enabled the rapid implementation of digital infrastructure, which would be challenging and time-consuming for a traditional government department.

Due to its commercial nature, Landgate must source alternative income to manage cash flows and maintain cost-effective operations, given public reliance on its services. Consequently, at times, development initiatives for key digital projects have been halted due to insufficient funding (Gainer, 2017).

The establishment of Landgate by the Western Australian Government has facilitated the development and digitalisation of key land management services. By using a statutory authority with a commercial management structure, the state government has been able to implement leading technologies with legislative support from the *Land Information Act 2006*. This case study underscores the need for governments to establish statutory authorities in sectors facing significant technological disruption and highlights the importance of conducive legislation for these authorities to not only function effectively but also regularly review their operations. Landgate provides a strong foundation for implementing and regulating AI and other digital technologies in the public sector across other Australian states and territories.

Property Valuation Services Corporation, Nova Scotia, Canada

Overview

Property Valuation Services Corporation (PVSC) is an independent, not-for-profit organisation that undertakes a mass appraisal to provide annual statutory assessments for over 645,000 properties across the 49 municipalities in Nova Scotia, Canada (PVSC, n.d.). Since 2019–2020, Nova Scotia, through the PVSC, has become the first jurisdiction in Canada to use AI to predict and apply property values for official assessments. Since 2018, PVSC has assembled a team of valuers, data scientists and software developers to integrate new ML software and algorithms that can synergise with the existing company operations (IAAO Artificial Intelligence Task Force, 2022).

Enabling policy and regulatory environments

The Government of Nova Scotia established the PVSC in April 2007 following the enactment of the *Property Valuation Services Corporation Act 2006*. PVSC is a corporate entity whose shareholders are the 49 municipalities of Nova Scotia. PVSC's primary objective is to provide assessment and related property information services for municipalities and the province of Nova Scotia. The corporation is managed by a board of nine to eleven people, including municipal councillors, independent members, past municipal staff and the Executive Director of the Nova Scotia Federation of Municipalities ('Property Valuation Services Corporation Act', 2006). The PVSC is given the statutory power to annually conduct mass appraisals of properties throughout the province through the *Nova Scotia Assessment Act* (PVSC, n.d.).

Since the programme's inception, the PVSC has worked alongside the Department of Statistics at Dalhousie University in Halifax, Nova Scotia. They have focused on the types of data required for the algorithms developed for property valuation, allowing for early adoption of ML in the data analysis process (IAAO Artificial Intelligence Task Force, 2022). The PVSC has also successfully worked with local universities to develop a cutting-edge method of mass property assessments. The development of a valuation corporation under the *Property Valuation Services Corporation Act 2006* has enabled the development of ML and AI applications in official assessments.

Outcomes and lessons

PVSC has been at the forefront of employing AI and ML to provide annual statutory assessments for the municipalities of Nova Scotia. The principal benefit of establishing such an independent corporation is its ability to work with local universities to develop strong data analysis and algorithms. Since implementing AI and ML models, PVSC has significantly increased its efficiency in its valuation processes. AI-based valuations now require 85% fewer in-person work hours each year while accurately conducting and explaining valuation assessments compared to traditional methods (IAAO Artificial Intelligence Task Force, 2022).

Despite obstacles during the COVID-19 pandemic, the implementation of AI and ML encountered limited software-related issues. Major obstacles to developing AI and ML models were public concerns about relying on computer-based decisions for statutory assessments (IAAO Artificial Intelligence Task Force, 2022). Most research and development for this programme occurred between 2018 and 2020, with no revenue generated, as it was conducted outside the annual statutory reporting operations of PVSC. The first trial, encompassing 30% of samples, was completed during the 2020 reporting period, approximately two years after the project's commencement. The first full integration of the software was planned for 2022, about four years after the project's initiation (IAAO Artificial Intelligence Task Force, 2022).

Overall, the development of AI- and ML-driven appraisal software has provided a significant upside to the ongoing operations of the PVSC. Institutional and legislative support of the *Property Valuation Services Corporation Act*

2006 and the *Nova Scotia Assessment Act* has enabled rapid implementation of the technologies and public trust by promoting transparency and accountability in AI- and ML-driven assessments. Additionally, partnerships with local universities played a crucial role, enabling research and development on algorithms and tools to analyse large local datasets. These partnerships not only facilitated the development of the AI and ML systems for statutory valuation but also led to cost efficiency compared to typical in-house technology development operations.

Analysis

A key theme emerging from both case studies is the effectiveness of having legislative instruments for separate corporate entities. These instruments have not only established these entities but also provided them with the statutory power to undertake a range of valuation and other land-based administration services.

Both entities have incorporated a commercial governance structure with a corporate board of directors, thus allowing autonomy and agile decisions across the organisations throughout AI and ML technology implementation. Having a corporate organisational structure has also facilitated effective funding arrangements, along with opportunities for investment in technology startups. For instance, Landgate provided early-stage funding and development assistance for PEXA (previously Property Exchange), leading to the successful launch of one of the first digital property transaction software in Australia in 2014 (Gainer, 2017). Another notable investment was made in ADVARA, a cloud-based software for digital land title registrations.

Landgate's investment in ADVARA also played a crucial role in the development of its own NLR. Since its establishment, Landgate has focused on streamlining and digitising the land sales registration processes. The development of key technologies such as NLR in 2015 has significantly improved the efficiency of property title registrations and transactions for Western Australia. Additionally, through the establishment of an innovation hub and grant programmes such as the SPUR Location Grants that seek to utilise the state's location information to benefit the land sector, Landgate has been ahead of the curve with technological developments.

Meanwhile, PVSC has strategically worked alongside the Department of Statistics at Dalhousie University and local universities to develop and maintain the databases and models for its AVMs (IAAO Artificial Intelligence Task Force, 2022). This case study demonstrates the upside of partnerships throughout the development and implementation of AI and ML valuation tools (Gainer, 2017).

Both case studies highlight the need for government agencies to reassess the structure and legislative power of departments that undertake statutory valuations, as they have, to date, not been conducive to the implementation and development of new technologies for digitally driven valuation (Victorian Government, 2024). Through the use of separate legislated corporate entities, PVSC and Landgate have successfully developed, adapted and most importantly matured alongside AI and ML technologies while still maintaining their statutory responsibility for valuation services.

With two case studies outlined and analysed to illustrate how AI and ML have been implemented successfully, the next section concludes by drawing key points and recommendations from all prior discussions.

Conclusion

Valuation with artificial intelligence and machine learning

This chapter has outlined the development and significant potential of AI and ML in property and real asset valuation. With considerable potential to assist in achieving SDGs 9 and 11, AI- and ML-driven valuation is a key tool decision-makers can employ to drive their jurisdictions' industry, innovation and infrastructure and create more sustainable cities and communities. While these technologies can enhance efficiency and potentially improve the accuracy of the valuation process and outcomes, the lack of 'human touch' required in the valuation of properties with a substantial degree of complexity is a major issue. Lack of transparency and the need for 'unbiased' datasets are also key factors limiting the reliability and quality of AI- and ML-driven valuation tools. Therefore, to maximise the potential of AI- and ML-driven valuation tools while minimising their shortfalls, it is imperative that conducive policy and regulatory environments be established. The two case studies of Landgate in Western Australia and the Property Valuation Services Corporation in Canada reinforce the importance of such environments.

Recommendations for agencies and readers

The case study of Landgate and its legislative framework illustrates a model that could be emulated in other Australian states and territories. State- or territory-level legislation similar to Western Australia's *Land Information Authority Act 2006* should be established to drive statewide adoption of AI and ML in mass appraisals for statutory valuation. Such legislation should seek to establish a statutory agency and provide it with the legislative power for technologically driven statutory valuation. Where possible, such an agency should be operated with a commercial management structure to expedite the process of developing and implementing AI and ML in its valuation methodology.

To achieve a fair market valuation of properties while aiming to create and maintain public and stakeholder trust in technologically driven valuation tools, transparency and reliability of algorithms and datasets must be ensured.

State- or federal-level legislation can be introduced or revised to ensure sufficient ethical standards are undertaken in the development and implementation of AI- and ML-driven valuation tools across both the public and private sectors. Legislation should also stipulate that their developers must fully disclose algorithms and variables employed in their AI- and ML-driven valuation models, including a clear rationale behind their usage. Such a requirement will assist in enabling interpretations of the tools, thus increasing stakeholder trust and potentially promoting wider adoption. Legislative instruments could also require periodic reviews and updates of the algorithms used and regular 'audits' of value estimates to improve the reliability and accuracy of the tools.

The PVSC case study underscores the importance of partnerships between statutory bodies and local universities in developing tailored datasets and algorithms required for AI- and ML-driven valuation models. Therefore, in addition to the legislation above, government agencies should seek to develop cross-sectoral partnerships between universities and developers of these tools through various incentives and mechanisms. Some of these could include grant funding, legislative requirements and outreach programmes.

Regardless of the mechanisms employed, agencies should use the appropriate instruments available to them to maximise the advantages (efficiency, reduced susceptibility to human bias and multi-sectoral application) and mitigate the disadvantages (the 'black box' nature, dataset bias, and lack of 'human' perceptions) of AI- and ML-driven valuation tools identified in this chapter.

The following are key questions for further deliberation by the readers:

1. Can – and should – AI and ML completely replace traditional property and asset valuation methods? Why or why not?
2. Is the valuation profession at risk from the rapid acceleration of AI and ML capabilities? Why or why not?
3. How can AI and ML technologies be used to simplify and streamline the workflow of property valuers?
4. How can regulations ensure ethical standards and minimise bias in the datasets and algorithms used in the development of AI- and ML-driven valuation tools?

Recommended resources for further reading

Gainer, M. (2017). *Embracing Disruption: Transforming Western Australia's Land Agency, 2007–2017*. Innovations for Successful Societies.

Knight, A. (2022). *Automated Valuation Models (AVMs): Implications for the Profession and Their Clients*. RICS.

Ota, A., & Uto, M. (2024). Variation in Property Valuations Conducted by Artificial Intelligence in Japan: A Viewpoint of User's Perspective. *Real Estate*, 1(3), 252–266.

References

Abidoye, R., Ma, J., & Lee, C. L. (2021). Barriers, Drivers and Prospects of the Adoption of Artificial Intelligence Property Valuation Methods in Practice. *Pacific Rim Property Research Journal, 27*(2), 89–106. https://doi.org/10.1080/14445921.2021.2001724

Aldoseri, A., Al-Khalifa, K. N., & Hamouda, A. M. (2024). AI-Powered Innovation in Digital Transformation: Key Pillars and Industry Impact. *Sustainability, 16*(5), 1790. https://www.mdpi.com/2071-1050/16/5/1790

API. (2024). *Rules of Professional Conduct*. Australian Property Institute Limited.

Australian Government. (2024). Australia's AI Ethics Principles. https://bit.ly/4fWIF6X

Baum, A., Graham, L., & Xiong, Q. (2021). *The Future of Automated Real Estate Valuations (AVMs)*. Said Business School. https://bit.ly/4fPOYsc

Biometric Information Privacy Act, (2008). https://bit.ly/3ZEqqMN

Brown, S., & Engler, A. C. (2023). *Governing the Ascendancy of Automated Valuation Models: Regulating AI in Residential Property Valuation*. Brookings Institution.

Burton, T. (2023). *AI Property Tool Set to Revolutionise Valuation*. The Australian Financial Review. https://bit.ly/4eyzHKN

Cajias, M. (2021). Artificial Intelligence and Real Estate – Not Just an Evolution, a Real Game Changer! *Journal of Property Investment & Finance, 39*(1), 15–18. https://doi.org/10.1108/JPIF-06-2020-0063

Cheung, W., & Yiu, E. (2024). *Use of AI in Property Valuation Is on the Rise – But We Need Greater Transparency and Trust*. The Conversation. https://bit.ly/499Kgmc

Comunale, M., & Manera, A. (2024). The Economic Impacts and the Regulation of AI: A Review of the Academic Literature and Policy Actions - IMF Working Paper No. 2024/065. SSRN. https://bit.ly/4fVlnND

European Commission. (2021). Proposal for a Regulation Laying Down Harmonised Rules on Artificial Intelligence (Artificial Intelligence Act). https://bit.ly/3ZnB5dg

European Commission. (n.d.). Data Protection in the EU. https://bit.ly/3CRIDgT

Farrow, L., Haywood, A., & McHenry, D. (2022). *CDEI | AI Governance: Full Report*. United Kingdom Government.

Ferguson, C., He, J., & Kim, D. (2023). China's New Rules For Generative AI: An Emerging Regulatory Framework. https://bit.ly/3BcsEJU

Floridi, L., Cowls, J., Beltrametti, M., Chatila, R., Chazerand, P., Dignum, V., Luetge, C., Madelin, R., Pagallo, U., Rossi, F., Schafer, B., Valcke, P., & Vayena, E. (2018). AI4People – An Ethical Framework for a Good AI Society: Opportunities, Risks, Principles, and Recommendations. *Minds and Machines, 28*(4), 689–707. https://doi.org/10.1007/s11023-018-9482-5

Gainer, M. (2017). *Embracing Disruption: Transforming Western Australia's Land Agency, 2007–2017*. Innovations for Successful Societies.

Gloudemans, R., & Sanderson, P. (2021). The Potential of Artificial Intelligence in Property Assessment. *Journal of Property Tax Assessment & Administration, 18*(2), 87–106.

Glumac, B., & Des Rosiers, F. (2021). Practice Briefing – Automated Valuation Models (AVMs): Their Role, Their Advantages, and Their Limitations. *Journal of Property Investment & Finance, 39*(5), 481–491. https://doi.org/10.1108/JPIF-07-2020-0086

Government of Canada. (2023). The Artificial Intelligence and Data Act (AIDA) – Companion Document. https://bit.ly/3ZiuUaG

IAAO Artificial Intelligence Task Force. (2022). A Review of the Methods, Applications, and Challenges of Adopting Artificial Intelligence in the Property Assessment Office. *Journal of Property Tax Assessment & Administration*, *19*(1), https://doi.org/10.63642/1357-1419.1244

IBM. (2023). AI vs. Machine Learning vs. Deep Learning vs. Neural Networks: What's the Difference? https://ibm.co/48S4Es9

Infocomm Media Development Authority. (2024). Singapore Proposes Framework to Foster Trusted Generative AI Development. https://bit.ly/4geNQOj

ISO. (2022). *ISO/IEC 22989:2022: Information Technology – Artificial Intelligence – Artificial Intelligence Concepts and Terminology*. ISO.

IVSC. (2024). *International Valuations Standards*. IVSC.

Jaafar, N. S., Mohamad, J., & Ismail, S. (2021). Machine Learning for Property Price Prediction and Price Valuation: A Systematic Literature Review. *Planning Malaysia*, *19*(17). https://doi.org/10.21837/pm.v19i17.1018

Jafary, P., Shojaei, D., Rajabifard, A., & Ngo, T. (2022). *A Framework to Integrate BIM with Artificial Intelligence and Machine Learning-Based Property Valuation Methods 17th 3D GeoInfo Conference*, Sydney.

Jafary, P., Shojaei, D., Rajabifard, A., & Ngo, T. (2024). Automated Land Valuation Models: A Comparative Study of Four Machine Learning and Deep Learning Methods Based on a Comprehensive Range of Influential Factors. *Cities*, *151*, 105115. https://doi.org/10.1016/j.cities.2024.105115

Knight, A. (2022). *Automated Valuation Models (AVMs): Implications for the Profession and Their Clients*. RICS.

Kosinski, M. (2024). What You Need to Know About the CCPA Draft Rules on AI and Automated Decision-Making Technology. *IBM*. https://www.ibm.com/think/news/ccpa-ai-automation-regulations

KPMG. (2023). *The Future of Real Estate*. KPMG. https://bit.ly/3Aqw0sd

Landgate. (2024a). *Landgate Annual Report 2023-24*. Landgate.

Landgate. (2024b). Our Legislated Role. https://bit.ly/3ZB6GK6

Lescrauwaet, L., Wagner, H., Yoon, C., & Shukla, S. (2022). Adaptive Legal Frameworks and Economic Dynamics in Emerging Tech-nologies: Navigating the Intersection for Responsible Innovation. *Law and Economics*, *16*(3), 202–220.

New South Wales Government. (2024). *Artificial Intelligence in NSW Planning*. NSW Government,. https://www.planning.nsw.gov.au/assess-and-regulate/development-assessment/artificial-intelligence-in-nsw-planning

OECD. (2024a). OECD AI Principles Overview. https://bit.ly/49hUNvT

OECD. (2024b). *Recommendation of the Council on Artificial Intelligence*. OECD.

Ota, A., & Uto, M. (2024). Variation in Property Valuations Conducted by Artificial Intelligence in Japan: A Viewpoint of User's Perspective. *Real Estate*, *1*(3), 252–266. https://www.mdpi.com/2813-8090/1/3/13

Pagano, T. P., Loureiro, R. B., Lisboa, F. V. N., Peixoto, R. M., Guimarães, G. A. S., Cruz, G. O. R., Araujo, M. M., Santos, L. L., Cruz, M. A. S., Oliveira, E. L. S., Winkler, I., & Nascimento, E. G. S. (2023). Bias and Unfairness in Machine Learning Models: A Systematic Review on Datasets, Tools, Fairness Metrics, and Identification and Mitigation Methods. *Big Data and Cognitive Computing*, *7*(1), 15.

Pantanowitz, L., Hanna, M., Pantanowitz, J., Lennerz, J., Henricks, W. H., Shen, P., Quinn, B., Bennet, S., & Rashidi, H. H. (2024). Regulatory Aspects of Artificial Intelligence and Machine Learning. *Modern Pathology*, *37*(12). https://doi.org/10.1016/j.modpat.2024.100609

Pettit, C. (n.d.). *AI-Powered Property Valuation Tool for City Planning*. University of New South Wales,. https://www.unsw.edu.au/research/partner-with-us/case-studies/frontier-si

Property Valuation Services Corporation Act, § SNS 2006, c 19 (2006). https://bit.ly/3VkKdhR

PVSC. (n.d.). How We Assess Property. https://bit.ly/3OGYXUE

Rathore, M. M., Shah, S. A., Shukla, D., Bentafat, E., & Bakiras, S. (2021). The Role of AI, Machine Learning, and Big Data in Digital Twinning: A Systematic Literature Review, Challenges, and Opportunities. *IEEE access*, *9*, 32030–32052. https://doi.org/10.1109/ACCESS.2021.3060863

RICS. (2019). *Valuation and Sale Price*. RICS.

RICS. (2021). *Automated Valuation Models: Roadmap for RICS Members and Stakeholders (June 2021)*. RICS.

Stanford Institute for Human-Centered Artificial Intelligence. (2024). *Artificial Intelligence Index Report 2024*. Stanford University.

Su, T., Li, H., & An, Y. (2021). A BIM and Machine Learning Integration Framework for Automated Property Valuation. *Journal of Building Engineering*, *44*, 102636. https://doi.org/10.1016/j.jobe.2021.102636

Tchuente, D. (2024). Real Estate Automated Valuation Model with Explainable Artificial Intelligence Based on Shapley Values. *The Journal of Real Estate Finance and Economics.* https://doi.org/10.1007/s11146-024-09998-9

Topraklı, A. Y. (2024). AI-driven Valuation: A New Era for Real Estate Appraisal. *Journal of European real estate research.* https://doi.org/10.1108/JERER-04-2024-0031

United Nations. (2015). *Transforming our world: the 2030 Agenda for Sustainable Development A/RES/70/1*. United Nations. https://bit.ly/3AoSs4X

United Nations. (2024). Harnessing Artificial Intelligence for Sustainable Development Goals (SDGs). https://bit.ly/3VoJwE2

University of New South Wales. (n.d.). Value Australia – Sharpening our Land and Property Decisions with Artificial Intelligence. https://bit.ly/4fBWPbR

Victorian Government. (2024). *2024 Valuation Best Practice*. Victorian Government.

Western Australian Government. (2024). *Decarbonising Western Australia and Our Trading Partners: September 2024*. Western Australian Government.

Yanamala, A. K. Y., & Suryadevara, S. (2024). Navigating Data Protection Challenges in the Era of Artificial Intelligence: A Comprehensive Review. *Revista de Inteligencia Artificial en Medicina*, *15*(1), 113–146.

Chapter 4

Coevolution of artificial intelligence technology and the valuation profession

Hao Wu and Rita Wong

Introduction

This chapter investigates the relationship between artificial intelligence (AI) technology and the valuation profession as co-developing/evolving phenomena, since significant progress has been evident since the beginning of the 20th century, giving continuous transformation at local and global scales. This means that the valuation profession faces challenges and opportunities in both the market economy tradition and other jurisdictions and cultures. The rise of AI raises the concern of radical occupational substitution that exhibits a strong natural selection flavour. Our primary interest is to take a critical and practical view of the technology domain in valuation as the human intelligence-enabled emergence of AI technology impacts the valuation profession.

This chapter is *not* a detailed study of AI technological implications in valuation service or AI-specific valuation and analysis. Some of these are deeply linked to our discussion of the interaction of technological change, service product and process and principal-agent identity and their organisation. Our primary interest is the most recent iteration of the technology-valuation-profession trilogy, focusing on technological and organisational impacts. That is, it focuses on professional services as currently provided by human intelligence, acknowledging that certain processes and skills substitutions by AI will reflect service quality change or technical disruption. It contributes to a meaningful measure of service productivity and stakeholders' wellbeing in property systems, and particularly, the built environment context.

This chapter critically considers the role of AI technology in its dynamic relation and interaction with the future evolution of the property valuation profession. Firstly, it examines the concept of AI technology and the development of this technical domain, investigating its practical legitimacy and distinction from human intelligence, while remaining open to the possibility and tendency of this integration. This indicates AI's dual nature of being human labour substitution and capacity enhancement, suggesting new process-outcome uncertainty. The chapter then examines the valuation profession as a social organisation, fitting into specific socio-historic context(s). It considers

DOI: 10.1201/9781003539254-5

the impact of an industry-wide transformation, with the main constituents revolving around valuation services as labour processes versus the jointly determined value of valuation practice and the profession using AI. Whether AI technology augments or substitutes professional valuation services (i.e. value-enhancing or -destructing) is a very relevant domain for exploration.

Several aspects of valuation practice that this chapter pays attention to include valuation problem/issue identification, data collection and analysis, methodology, reasoning, judgement and communication. To contextualise, we reference one of the most vibrant and rapidly evolving regions, namely the Asia-Pacific region, and its professional practices. It approaches the region's current valuation practices and professionalism as well as focuses on emerging AI's knowledge benefit to improve/enhance process, outcome and professional competence. Further, it considers risks and opportunities for the valuation profession. The chapter envisions a mutually dependent relationship of human agency and artificial intelligence to enable a sustainable socio-technological culture for AI to allow co-production via its interaction/integration with the valuation profession.

Artificial intelligence technology

There exist at least two ways to interpret technology. One being a historic narrative of socio-technical progress/change, and the other being an anthropological view of technology, which is unsubstitutable on moral grounds. Technology-profession's relational role might be expressed as '... processes of technical-industrial production... "get" technology "intelligently" in hand...' (Heidegger 1977), which corresponds to the instrumental and anthropological concept of technology as a 'traditional' mode and an impression of AI technology. Like other technologies, AI, in the Heideggerian sense, is a way of revealing, and its self-revealing features represent what he would call 'enframing' that goes beyond its typical meaning.

As to the fundamental nature of AI technology, disagreement remains as to whether AI should only be derived from human intelligence (i.e. mimic human intelligence), or could it emerge from resource domains other than humans? What we mean by 'intelligence' and how we describe/define it plays a critical role in our practical choice of value and other determinations of a good, such as a property (Brody 1999, Pfeifer and Scheier 1999). And one needs to note that intelligence is culturally embodied, and its core value is context-specific (Sternberg 2004; Sternberg 2020).

Besides the essence and nature of technology, enframing and describing what has taken place in technological advancement directly links to its impact on labour cost, industrial structure and people/agents' wellbeing. Susskind and Susskind (2018, p. 129) offer a four-dimensional framing of technology change, namely, its exponential growth, increasing system capacity, pervasiveness in social life and increasing human connectedness. What this implies is that the impact of technology on an existing profession is complex and often uneven, with varying degrees of process-outcome uncertainty.

The transformative power of technology in driving societal change and productivity improvement is considered both fundamental and revolutionary. One of the most popular topics in recent years is the emergence of AI and its transformative power in society's practices, even though its definition remains ambiguous. Some prefer AI as machine technology that mimics human intelligence (Saghiri et al. 2022), while others argue that human intelligence is not necessarily a condition of AI (Susskind and Susskind 2016). 'Human-like' capabilities are a key ambition of its development, use and imagination. Nonetheless, every AI system so far emerged is compositional, with its capability constantly evolving (Moro-Visconti 2024).

The degree and significance of process-outcome uncertainty are sometimes treated as an important measure of the power of technological transformation in society or its sub-sections. The transformative power of new knowledge is usually represented as a co-development or co-evolutionary process (Schumpeter 1942; Nelson 2001; Arthur 2009). The process embeds the dual nature of AI in society, which triggers substitution and complementarity, which build collaboration. It is hoped that the transformation upgrades, hopefully not at the detrimental cost of existing labour arrangements and their institutions.

Hence, the relationship of human labour and its technological alternatives/substitutes can be clearly shown in the progression of AI-aided valuation. The value of the valuation profession is shown in the valuation process as a flow of valuation services delivered by an organisation of valuation experts. With specific discussions of AI's impact on operational efficiency and productivity of valuation services, processes and structure, it is argued that AI technology has existing and potential impact on every stage and component of valuation services and the structure of the valuation profession.

The property valuation profession

A labour servicing process

The idea that a profession represents division of labour in the production process of expert service is recognised in major studies of the social organisation of professions. 'Professionalism has been the main way of institutionalising expertise in industrialised countries'. (Abbott 1988, p. 323). Existing systems of profession emphasise expertise being possessed by an individual expert and put the specialised knowledge to practical use via certain organisational models. It mimics a specific case of the distribution and use of knowledge in society (Hayek 1945).

As for the defining features of a profession, four are shared across contemporary times: (1) specialist knowledge, (2) admission on credentials,

(3) professional regulation, (4) a common set of values (Susskind and Susskind 2015, p. 15). Valuation as a labour process (which we argue is what a valuation service is essentially about) represents expert service that is a purposeful means for productivity improvement. As to its historic origin, emergence and development, the valuation profession comprises at least some public good elements or features. As in most professional traditions, there are significant differences in the ideal of professionalism in terms of this public good spirit. If we consider the possession, use and institutionalisation of specialised knowledge, this professional spirit can be traced back to the ancient histories of various civilisations, nations and societies, even if their ways of practice and organisation vary.

As human productive labour and value creation from biological and cognitive capacities are being challenged by new technologies, a question emerges as to what new services human labour can outperform or co-exist with them? 'With the deregulation of the labour market, diversification of professional services, standardisation of professional work, multidisciplinarity and flexibility come into play'. (Haapakorpi 2022, p. 443 in Susskind and Susskind 2016, pp. 114–119). Current professions may expand or adjust their scope of service by developing a new service portfolio and by identifying new opportunities to assist their transformation (Haapakorpi 2022). This is because, compared to raising labour efficiency, technological reconfiguration embeds innovation value or a shorter-term comparative advantage. Gilbertson and Preston (2005) raised a similar concern about the future of professional valuation services. More recent examples include attempts such as developing data-driven valuation services.

Consider a profession powered by the spirit of professionalism as a system of service labour, products, and their organisations (e.g. valuer, firm, industry); ethical value bonds them to a specific cultural-institutional context. The system of professions and their associated industries produces service flows to allow the evolution of socio-organisational functionality. A profession represents a specific identity of organised human practical knowledge. For example, a profession represents a group of specialists/experts who hold or belong within a technological knowledge domain. One may see a profession as a model to organise human and non-human activities, which, from an identity perspective of trust, value and power, jointly determine the scope and content of expert service such as valuation. The same should apply to other professions such as medicine, law, engineering and architecture.

Some historic-contextual perspectives

The historical development of the valuation profession tradition in the family of human services is rooted in the industrial revolutions. One of the widely accepted/practised organisational approaches to the valuation profession is the territorial scales of nation, region/state, and local authority. What is under-considered is the diverse systems and contexts that govern the valuation profession (Gilbertson and Preston 2005). Some systems are primitive and/or dependent on traditional practices of cultural norms to provide expert service to facilitate communication and trade. In mature market systems, technology and industry practice are closely linked to the idea of profession and associated services and organisational structure. Over time, both valuation practice and the valuation profession adjust substantially.

It is raised that the perceived transformative power of AI technology will continuously alter existing structures of the valuation profession (Wilkinson et al. 2018a). Less obvious is the extent to which and in what ways or aspects the change will occur. It is also unclear what kind of new private and public demands will emerge and how these will be rebalanced. Moro-Visconti (2024) offers a detailed discussion of the meanings, means and applications of AI technology to valuation. His narrative does not discuss AI's interactive relationship with the valuation profession.

The continuous transformative effect of AI technology on the valuation profession should be based on the concern of, and impact on, its core social services. It is important to review our conception and interpretation of what the core value of professional services is. Of course, one can perceive the value(s) of the valuation profession from multiple angles. For example, Ruiz et al. (2008) suggest it should include the dimensions of service quality and equity, confidence, benefits and perceived sacrifice in a business marketing context. The same question(s) are raised and examined, in more depth and to a greater extent, in other major professions such as medicine, law, surveying and engineering. There is a small stream of literature on future conditions of the valuation profession in the Australian context (e.g. Gilbertson and Preston 2005; Elliot and Warren 2005; Wilkinson et al. 2018a, 2018b). We do not intend to conduct a comprehensive survey on these or extend the context much beyond the English tradition.

An interesting stream of inquiry concerns the likelihood and conditions of a 'post-professional' future (Susskind and Susskind 2016). Some urge major professions to rethink their practices in the context of the digital economy and community as 'post-industrial' and/or 'post-professional', such as the so-called professional portfolio career (Haapakorpi 2022). This may trigger labour insecurity concerns and multidisciplinary collaboration demands. One should also note that there are varying development stages of industrialisation around the world, and their corresponding paths of professionalisation can diverge substantially. Debate is raised and conducted mainly in the advanced regions and nations. Regions and jurisdictions where their priorities are development tend to put less emphasis on professional service transformation. Some adopt less industrial or market-based systems to facilitate resource allocation, production and service-knowledge communication.

Coevolution of AI technology and professional valuation services

Technic-labour coevolution

The idea of technic-social coevolution (or coadaptation) is not new. Janzen (1980) explains the concept and its applicability in a biological context. Nelson (2001) and Kallis (2007) interpret the concept in the context of evolutionary historical economics. More recently, a socio-technical perspective emerges, where the concept is adopted and applied to the area of changing relationships of technology and institutions (Zhang 2020) and AI-specific technic-social environments (Munoz 2024). In the context of the valuation profession, what is likely to emerge are new labour models of valuation services and industry dynamics.[1]

As AI technology progresses, it breaks existing collective organisational structures and reimagines how we produce, operate/manage and use knowledge and technical means. It is a socio-technical process, so it is important for us to be clear about the time scale we use to describe their relationship and interaction. Conceptually, technology exhibits its dual roles in transforming the profession.[2] The first is sustaining, enhancing and augmenting (such as machine-learning-powered automation). The second is innovating, disrupting and destructing (such as labour substitution). It is important to note that although these two roles and effects are conceptually distinctive, they form part of the same process and relationship of professional practice. This leads to a logical approach to measuring and assessing the impact of technology on a specialised knowledge field.

As AI socio-technical systems develop the capability to outperform and replace human services such as professional valuation services, will it discourage existing human effort? The answer is affirmative, which corresponds to the new technic-social and industrial transformation of work-life patterns. A follow-up challenge is 'what will motivate our painstaking effort to create and sustain the new meanings of work-life relation to improve human existence and societal progress?' As a result, the very meaning of personal development and progress may change. And it is envisaged that the community spirit of public value and people's identity may motivate productive activities. This means that people, being identified as 'consumers' or 'producers', need social meaning in life as a primary motive to sustain their actions (as Max Webber would put it).

A more pessimistic view, however, concerns both the significant relative decline of human computational capacity to AI progress (as Moore's Law associates) and the actual ability to control and possess decision power by human actors. The latter being the autonomy of human decision and professional advice. Logically, we may fall back into the technological black box that is increasingly out of touch, even for the profession. This means ideology and ethical culture context matter, if not explicitly, for the proper condition and functionality of the profession across socio-technical regimes. This impinges on fundamental redistribution and powerful externalisation of knowledge stock and its use in society. When put into this context, it may bring new prospects or debate as to Hayek's (1945) classic insight of the relative efficiency of institutions (e.g. free market vs. planned economy) concerning the use of knowledge in society. And it is unlikely that the valuation, or other major professions, will be untouched when the structure of professional knowledge no longer reflects its actual use and distribution.

While considering the powerful and profound transformative effect that AI technology induces, it is critically important to maintain the quality and development of human physical and mental capabilities and their knowledge stock. It is envisaged that AI technology will trigger and enable new professional services and career opportunities. The (re)stratification of the profession will show the restructuring effect of its creative and destructive power. This also envisions the emergence of new competencies, a new configuration of professional services, and potentially a new organisational structure (or oversight) of the profession.

AI technology enabled valuation professional services

We now pay attention to the valuation-specific process, considering the co-evolving nature of technology and profession. Central to the process is a new understanding of human and machine capacities integration in AI-aided valuation. It is based on a critical view of intertwined forces of creative destruction (Schumpeter 1942) and the deliberation of expertise from craft to commons (Susskind and Susskind 2015). These represent new forms of production and distribution of expert knowledge. It is the organisation of valuation experts to conduct AI-aided valuation as 'joint production' processes.

Understanding, use or impact of AI-aided valuation

Practically, should the traditional algorithm, machine learning, and artificial intelligence-based 'technologies' be distinguished in valuation practice? At the time of writing, machine learning-based algorithms with data-driven analytics are experiencing significant progress, while the idea of 'intelligence' for professional valuation faces high-level uncertainty (Szumilo and Wiegelmann 2024). Examples of major recent progress in valuation services are largely in mass valuation and homogeneous product sectors. Like the AI-aided practice in tax law, the statutory and rating valuation sector has the required infrastructure to support AI practices.[3] Applications also spread to commercial real estate, especially logistic property and portfolio analysis, and have been powerfully transforming these sectors (JLL 2025; Leow and Lindenthal 2025). Which brings up the central question: what is the fundamental need for AI-aided valuation in future economic affairs?

The main comparative advantage of AI to the valuation profession is multifaceted. Many have been explored or used in practice. It has been shown to bring scale, speed and some quality advantages to allow processing mass data quickly and accurately, reducing some human errors/biases and removing human mundane/repetitive tasks to allow time for innovative and thought-provoking activities (Kuppan et al. 2024). What underpins it is the high likelihood of the quality of the valuer being elevated, implying tension and a tendency to upgrade the capacity of the valuation process and the profession. Reviewing the history of industrialisation, this upgrade means shifting the labour force via technology-labour substitution. It has also enhanced professional insights through advanced use of data for real pattern discovery, trend recognition and prediction.

Technological progress, as knowledge revealing/enframing, increases process efficiency and productivity while altering service performance (product quality) through streamlined task processes for valuations. This will accelerate business pricing and operations in a competitive market environment. Again, that trade-off between process efficiency versus service quality remains a critical issue. Accompanying it is the concern of a significant fee cut without a 'cost' decline. This implies labour cost drops through labour-technique substitution, while transaction (e.g. learning and management) cost is likely to rise, or it will remain high. AI-aided valuation enhances risk assessment, which is achieved through real-time data analytics, continuous risk monitoring and policy intervention. AI-aided valuation being culturally dependent also poses a challenge (e.g. bias) from data narrative or other qualitative concerns. For example, design, historic context and sentiment correspond to local seller/buyer behaviour or broader system-wide natural, social or biological shocks such as major tsunamis, earthquakes, pandemics or wars.

AI applications and their integration in valuation services

One of the popular views of valuation experts is that AI is only as powerful as the algorithm that it is 'wired' with, plus the quality of the information (datasets) it is fed. This is an indication of the core components of the new professional capacity and expertise, which may help determine the future composition of the valuation workforce. Actual transformation and integration are unlikely to be a straightforward (or smooth) process. An AI technique expert can write the programme, but will need professional instructions and interpretation – something that the basic principle of division of labour underlines. Advanced technology, such as AI, improves knowledge transference and sharing by institutionalising it among people and organisations, but the challenge of knowledge distribution in human society remains.

The shared nature and scale effect of combining information and knowledge also mean major managerial challenges to achieve and maintain the quality and efficiency of valuation. Property valuation features highly regulated markets for property rights and information, which require high management costs. As we see in the context of Creative Commons, new governance systems may emerge to enable AI-aided valuation services. The valuer's recent observation and experience from the Asian-Pacific region suggest that some transformative shift is occurring primarily in the mass residential sector and, to a lesser extent, in rating, portfolio asset and logistic industrial practices. Although the joint-production and co-evolving features are present, whether it represents a paradigm change remains unclear.

Risk and compliance of AI predictive analytics in valuation

One of the valuation facts of professional practice is that in fast-falling markets, be it actual or market sentiment, valuation adjustments often lag, and vice versa in a rapidly rising market. The profession faces market cycles of periodic under- and over-supply of market evidence, which leads to bias and inaccuracy for market valuation practice (Whipple 2006). In this regard, AI-aided valuation may help augment the profession's ability to predict market value during under-/over-supply of reliable market evidence.

The concern of professional conduct of AI-aided valuation is subject to continuous debate as it deals with the important idea of professionally responsible AI (Hatfield 2019). Important to note is that this has not touched on the ethical conduct of human/non-human general AI. Although advanced machine languages emerged, making non-human communication closer to realisation, human-like machine thinking appears to be an innovation for a distant future. Currently, the bias and accuracy of AI-aided valuations concern how the systems and algorithms are programmed. The transparency concern as to how AI-aided valuation is prepared and interpreted ensures significant insurance and communication costs are imposed on valuers, AI specialists, users and other stakeholders.

Considering professional accountability and responsibility, the valuation profession has yet to initiate substantial proposals to address joint production (or integration) responsibilities of human and machine where output is generated by a machine designed by a human who controls and manages it.[4] The organisational nature of the valuation profession also expects their governing bodies, such as professional institutes, to use their ability to endorse appropriate development and use of AI socio-technical systems on behalf of their members.

As technological progress within 'private markets' usually outpaces its regulatory progress, the 'soft ground' concerning the legitimacy and reliability of AI-aided valuations affects its effective use in a few key areas/parties in industry practice. For example, it affects the willingness and confidence of financial institutions to lend on AI-based valuations. Given the weak regulatory environment or institutional vacuum in some cases of AI-powered practices, regulatory compliance costs can be extremely high. Understandably, professional indemnity insurance for 'black box valuation' can be exceedingly costly to sustain the proper functionality of private valuation practices. These all point to major concerns of proper governance and ethical standards, which are discussed later in the chapter.

Legal and ethical implications of AI-aided valuation

An emerging area of professional and academic interest is the legal and ethical implications surrounding data and cybersecurity (Horvitz and Mulligan 2015). Professionalism assumes valuers treat client data confidentiality as of paramount importance. Given the semi-open and shared nature of data in AI-aided valuation, it is hard to be certain as to what levels of data protection need to be in place. These legal and ethical concerns of data privacy and cybersecurity bring practical challenges to the valuer because managing the complex information network is virtually an impossibility for them. Therefore, in the era of AI-aided valuation, the boundary and scope of a valuer's legal/ethical conduct and professionalism are likely to shift.

The likely boundary shift of professional services also raises concern about valuers' roles in AI-aided valuation. It is envisaged that a human valuer still plays a crucial role in valuation as new AI technology integrates into that process. For example, a valuer is responsible for building, training and feeding the AI tool, as they are responsible for interpreting outputs and validating them. They are responsible for contextual relevance of valuation, such as market sentiments and intangible property features, which the qualitative AI tools are unable to capture well. The valuer is also responsible for client communications – the entire process from enquiry to delivery and reporting. For as long as the client is human, the valuer has a duty of care to ensure and reinforce quality and trust in their services.

The rapid development and use of AI technology also has a profound effect on dividing the profession based on generations of valuers' varied levels of technology literacy. Filling this gap demands continuous learning and education, which is challenging due to competition for knowledge workers. In other words, 'talent is tight'. Clearly, it is the joint responsibility of industry bodies, government, clients and educational institutions to ensure valuers are appropriately trained and supported to keep up with technical progress. It needs to properly treat the question of how to best deploy varying levels of AI skills to valuation tasks.

Global perspectives and cross-cultural challenges in valuation processes and practices

Since the rise of network society and globalisation, the notion of 'think global, work local' is increasingly a new norm for the valuation profession. Although property value is highly context and location-dependent, the execution and management of property transactions is increasingly globally driven. As AI is a global socio-technical phenomenon, its use in valuation will benefit from international best practices and policies. That is, international practice and valuation standards are in line with AI-aided valuation practices.

The rise of geopolitical conflict and politico-economic instability raises global uncertainty and concerns of periodic switches from globalisation to localisation, led by the world's superpowers. As these global patterns will have implications on economic systems and property markets, their relationship with professional practice is complex. As countries internalise, valuers tend to respond to the trend by focusing on local market dynamics. This includes local demand (incl. historic trend), infrastructure and policies. This may continue and reinforce AI's assistance in augmenting professional services as well as its integration or convergence within human-based practice. The same applies to cost and return on investment concerns of small professional firms in incorporating AI-aided valuation. Apart from the efficiency and profitability of valuation services, clients of valuation would play an increasingly critical role in the development and conduct of AI-aided valuation.

Considering the logic and structure of capital at global to local levels, property valuation services and their clientele are largely institutionalised, gearing towards the demand of major lenders, public sectors and regulatory bodies. Substantial and ongoing demand for valuation services at a global scale is derived from financial reporting and financing requirements. Organised capital means that client/market perception and preference for AI-aided valuations powerfully, if not decisively, impact the valuation profession. Given the co-evolving nature of AI-aided valuation practices, it is essential to allow close collaboration of academia (R&D sector), government and industry/market entities to retain and strengthen their relevance and expertise through continuous innovation, development, training and education and learning by doing.

Governance of an AI-enabled valuation profession

A vision of the impact of digital technology on valuation services and stakeholder value of valuation services is an important way to reflect on how society transforms. This reveals structural change regarding existing and alternative policies and standards. The last few decades have seen the rise of industry self-regulation under national and international regulatory boundaries. Some bodies have developed and adopted more centrally administered, often top-down, specialised service systems. For a rising trend of self-employment and small-scale independent professional services, their regulation thereof becomes more decentralised with a parallel move to statutory homogeneity. This is because standardisation is usually a means to reduce social costs, especially enforcement and managing diverse interests. As the validation cost in the digital era of decentralised knowledge rapidly rises, the relative benefit of standardisation also rises, suggesting an underlying process of industry and market transformation.

The coevolution proposition helps put relevant policy and organisational codes of conduct into context: much is on human development and (professional) capability building and enhancement. Both valuation professions and society at large potentially realise the lag of an institutional framework to govern AI-aided valuations and ethical conduct, such as data security and privacy. As such, it is critical to repeatedly raise such questions as what is the role of professional standards, such as codes of conduct and best practice guidelines, in the era of digital transformation and AI-aided professional services?

The notion of institutionalised professional knowledge is an important one, particularly from the perspective that its scope is powerfully challenged by informal, flexible, decentralised practical knowledge in the era of AI technology and associated systems and products (Benkler 2002). The question of what qualifies as being professionally responsible AI technology and how to design/organise it is as relevant to older/established professions such as law and medicine as to more modern ones such as valuation.

Our discussion also brings into consideration the so-called (post)professional practices in a post-regulation era. The reason for this imaginative view is due to a rising social realisation that the nation-state governance of shared knowledge and expertise as common resources requires effective support from non-governmental sources. In such jurisdictions, where a profession and the services it provides have been self-governed through and within a broader regulatory system, new AI-enabled valuation practice shall have its corresponding governance systems. And we would imagine each has its composition of regulatory, production and community bodies, including, importantly, training and education bodies as the place and institution for generating, testing and sharing basic and practical knowledge that is AI-enabled.

Concluding remarks

Throughout human history, the evolving relationship of human and non-human beings has been a transformative process. The latest wave, being the coevolution of knowledge and the human bio-system, can be argued as a fourth industrial revolution (Schwab 2016). In an increasingly globalised network context, one should recognise that multiple systems of valuation and other professions co-exist. All are interacting with emerging technologies and within their socio-technic environments. This chapter seeks to contribute insights and ideas to support the sustained functionality and productivity of the valuation profession. It seeks to envision a collaborative relationship of human and artificial intelligence by enabling a sustainable socio-technical process for AI and its continuous interaction/integration with the coordination and governance of existing institutions, such as the valuation profession.

Valuation expertise represents a form of specialisation that carries the ethics of public spirit within its professional power. It is largely a service to facilitate market exchange and associated production processes, and what has traditionally made them special reveals what is behind it: specialised knowledge and division of labour. AI technology simultaneously transforms the client, product and governance sides of the valuation profession, locally and globally. The impact is through changing the relative distributive efficiency in knowledge productivity. We argue that uncertainty remains high on whether AI technology will improve both, one or none of these aspects. And if it does, to what extent will it impact existing industry participants and stakeholders? This will be of significant policy relevance.

The relationship of valuer and AI may be described as co-evolutionary, being an innovative process of socio-technical transformation. As anthropological and technological evolutions are continuous processes, we may adopt a 'stages approach' and tie it back to the relevance of the market to a specific profession, such as the valuation or surveying profession. We may take a cross-section (e.g. geographic stance) of the varying AI-aided valuation practices due to the differences and unevenness of their global distribution, affected by developmental and institutional conditions across territories, jurisdictions and societal cultures. Like the argued insights on technology's impact on professions (Susskind and Susskind 2016, Haapakorpi 2022), the transformative effect of AI technology on the valuation profession will be incremental and potentially lead to fundamental structural change in its production mode. The current stage appears to enhance and realise values through demand for labour services. The possible future stage(s) of labour service substitution could be market-driven, such as capital influence; government-driven, such as administrative; or community-driven, such as cooperative, valuation systems.

In the context of professional services, what is being upgraded or substituted is process efficiency. What can/will be enhanced is communicative and decision efficiency. It is argued that AI technology is unable to directly substitute human judgement, such as critical thinking. As Elon Musk once argued, there is a need for people to focus on first principles to know the role and impact of technology and then to apply them innovatively. This applies to socio-technical practices such as valuation. Constructive education and effective training ensure solid input and strategic upgrades of the profession's knowledge. As the historic processes of human-machine interaction can be a co-evolutionary and/or mechanistic causal relationship, the future will be dependent upon our choice of their co-production modes and their governing regimes. This dynamic relationship will shape the future role(s) of AI technology in valuation services and the structure and conduct of the valuation profession.

Discussion questions

Discuss the concept and domain of artificial intelligence technology and its relevance to the core practice of property valuation.

Describe the concept of profession and its main features. Discuss the possible relationship between AI technology and the valuation profession.

Notes

1 The notions of profession and professional service are essentially social constructs. As technology is socially embedded and historically dependent, the valuation profession (a labour organisation domain) and AI (an information technology domain) represent an interconnected or co-evolving relationship.
2 The nature of technology is reflected by its ability, its dual roles, to improve/enhance human capacity as new knowledge is being incorporated, or to risk/substitute human capacity due to its superior knowledge power to replace inefficiency in an existing system.
3 For example, the platform systems that generate valuations and the systems that facilitate the internal workflow of valuation services. Or as virtual reality and other human-machine interactive tools are used for online viewing of the premises. In China, Beike and Yuguijia are two examples. In Hong Kong, HSBC and Bank of China use AI-aided platforms to support mortgage lending, while major agencies (Centraline and Midland) have adopted AI-based algorithms. Japan's Sony Real Estate and Yahoo Japan, and Australia's CoreLogic and Jones Lang LaSalle have AI-aided appraisal/valuation tools and services.
4 A recent attempt in the context of the legal (taxation and accountancy) profession is Hatfield (2019). He argues that, like professional peer review, one's professional responsibility is their willingness and ability to judge when it is reasonable to rely upon AI-aided analysis and recommendation. This means trust and accountability.

References

Abbott, A. (1988). *The System of Professions: An Essay on the Division of Expert Labour*. University of Chicago Press, Chicago.

Arthur, B. W. (2009). *The Nature of Technology*. The MIT Press, Boston.

Benkler, Y. (2002). Coase's penguin, or, Linux and 'the nature of the firm'. *Yale law journal*, 4(3) 369–446.

Brody, N. (1999). What is intelligence? *International Review of Psychiatry*, 11, 19–25.

Elliott, P., and Warren, C. (2005). The valuation pro,fession in Australia. *Australian Property Journal*, 38(5), 362–367.

Gilbertson, B., and Preston, D. (2005). A vision for valuation. *Journal of Property Investment & Finance.* 23(2), 123–140.

Haapakorpi, A. (2022). The quality of a professional portfolio career in the post-industrial era. *Journal of Education and Work*, 35(4), 441–454.

Hatfield, M. (2019). Professionally responsible artificial intelligence. *Arizona State Law Journal*, 51, 1057–1122.

Hayek, F. (1945). The use of knowledge in society. *The American Economic Review*, 35(4), 519–530.

Heidegger, M. (1977). The question concerning technology. In *Basic Writings*, Harper Collins, New York.

Horvitz, E., and Mulligan, D. (2015). Data, privacy, and the greater good. *Science*, 349(6245), 253–255.

Janzen, D. H. (1980). When is it coevolution? *Evolution*, 34(3), 611–612.

JLL. (2025). Artificial intelligence: Real estate revolution or evolution? Retrieved October 11, 2025, from https://www.jll.com/en-au/insights/artificial-intelligence-and-its-implications-for-real-estate.

Kallis, G. (2007). When is it coevolution? *Ecological Economics*, 62, 1–6.

Kuppan, K., Bhaskar, A. D., and Divya, B. (2024). Foundational AI in insurance and real estate: a survey of applications, challenges, and future directions. *IEEE Access*, 12, 181282–181302

Leow, K., and Lindenthal, T. (2025). Enhancing real estate investment trust return forecasts using machine learning. *Real Estate Economics*, *53*(3), 1–33.

Moro-Visconti, R. (2024). *Artificial Intelligence Valuation*. Palgrave, New York.

Munoz, F. (2024). The coevolution of technology, markets, and culture: the challenging case of AI. *Review of Evolutionary Political Economy.* 5, 511–533.

Nelson, R. (2001). The coevolution of technology and institutions as the driver of economic growth. In *Frontiers of Evolutionary Economics – Competition, Self-Organisation and Innovation Policy*, Foster, J. and Metcalfe, J. (eds), Ch2, 19–30, Edward Elgar, New York.

Pfeifer, R., and Scheier, C. (1999). *Understanding Intelligence*. MIT Press, Cambridge.

Ruiz, D., Gremler, D., Washburn, J., and Carrion, G. (2008). Service value revisited: specifying a higher-order, formative measure. *Journal of Business Research*, 61, 1278–1291.

Saghiri, A. M., Vahidipour, S. M., Jabbarpour, M. R., Sookhak, M., and Forestiero, A. (2022). A Survey of Artificial Intelligence Challenges: Analyzing the Definitions, Relationships, and Evolutions. *Applied Sciences*, *12*(8), 4054. https://doi.org/10.3390/app12084054

Schumpeter, J. (1942). *Capitalism, Socialism and Democracy*. Harper & Brothers, New York.

Schwab, K. (2016). *The Fourth Industrial Revolution*. Currency, New York.

Sternberg, R. (2004). Culture and intelligence. *American Psychologist*, 59(5), 325–338.

Sternberg, R. (2020). *The Cambridge Handbook of Intelligence*, 2nd edition. Cambridge University Press, Cambridge.

Susskind, R., and Susskind, D. (2015). *The future of the professions: How technology will transform the work of human experts*. Oxford University Press, Oxford

Susskind, D. and Susskind, R. (2016). *The Future of the Professions: How Technology Will Transform the Work of Human Experts*. Oxford University Press, Oxford.

Susskind, D., and Susskind, R. (2018). The Future of the Professions. *Proceedings of the American Philosophical Society*, 162(2), 125–138.

Szumilo, N., and Wiegelmann, T. (2024). AI: real estate's new roommate – the good, the bad and the algorithmic. *Journal of Property Investment & Finance*, 42(2), 211–217.

Whipple, T. (2006). *Property Valuation and Analysis*, 2nd edition. Lawbook Co, Sydney.

Wilkinson, S., Antoniades, H., and Halvitigala, D. (2018a). The future of the Australian valuation profession. *Property Management*, 36(3), 333–344.

Wilkinson, S., Halvitigala, D., and Antoniades, H. (2018b). Educators, professional bodies, and the future of the valuation profession. *Property Management*, 36(4), 389–399.

Zhang, Y. (2020). The map is not the territory: coevolution of technology and institution for a sustainable future. *Current Opinion in Environmental Sustainability*, 45, 56–68.

Chapter 5

Challenges and opportunities of data collection and computational analysis using AI in real estate valuation

Xin Janet Ge and Peter Raptis

Introduction

Traditionally, property valuation has relied heavily on personal assessment of properties based on comparable sales and lettings, physical inspections and analysing local market conditions. Property valuers often relied on a combination of professional networks to source information that was not always available in the public domain. This method was time-consuming and often subject to human error or bias (Zurada et al., 2006). Given a convergence of technological advancements, data availability, and a need for more efficient valuation processes, Artificial Intelligence (AI), which is the ability of a computer to mimic human thought, is now disrupting the valuation process and profession. AI applications in property valuation have emerged and have been rapidly adopted in residential real estate, but less so in the commercial property industry. The rise of big data in real estate—encompassing transaction records, market trends, demographic information and geographic data—has provided a wealth of information that AI technologies can analyse.

AI can assist in real estate valuation through enhancing Automated Valuation Models (AVMs), which estimate property values using data like recent sales of similar properties, property features, location trends and market conditions. Machine learning (ML) is a way for computers to learn from data. As machine learning techniques have evolved, they have become capable of analysing complex datasets to identify patterns and predict future outcomes. ML algorithms can now process vast amounts of data to estimate property values more quickly and accurately than traditional methods (Wei et al., 2022). AI can automate repetitive tasks in the valuation process, predominantly in residential, such as data collection and preliminary analysis. This automation not only speeds up the valuation process but also allows for more frequent updates, reflecting real-time market changes. AI models can analyse numerous variables simultaneously, reducing the likelihood of overlooking critical factors that influence property values. This can lead to more reliable valuations and eliminate bias, which is particularly beneficial for lenders, investors and buyers.

DOI: 10.1201/9781003539254-6

AI can be used in various applications, including automated valuation models (AVMs), risk assessment tools for lenders (Krivorotov, and LaCour-Little, 2021; Liljegren et al., 2024), market trend analysis (Krivorotov et al., 2020), and property investment analysis (Nguyen, 2019). These applications help stakeholders make informed decisions based on predictive analytics. Though AVMs are not yet recognised as formal valuations, with more and more photo imagery being available for properties, AI computer vision models can now more accurately assess the condition of the property. This reduces the reliance on physical inspection of valuers. As AI becomes more prevalent in property valuation, regulatory bodies are beginning to establish guidelines to ensure transparency, fairness, and accountability in automated valuations (Cheung, 2024), addressing concerns about bias and accuracy. Integrating AI into property valuation represents a significant shift towards data-driven decision-making in the real estate sector, aiming to enhance efficiency, accuracy and accessibility.

One of the main challenges of AI in residential real estate is simulating human buying and selling decisions and patterns. There is considerable subjectivity and psychology in residential property markets, which AI has previously not factored into the models. However, this is now changing; for example, factors such as seasonality (spring vs winter) and personal decisions, such as proximity to schools, hospitals and even family members/community, are included. Commercial property, on the other hand, is not as subjective as residential property and is, therefore, ripe for disruption by AI. Given the challenges of AI application in property valuation, this research paper aims to understand valuers' perceptions of applying AI in property valuation. The research investigates AI in property valuation, particularly emphasising data collection in the context of leveraging AI in valuing different types of real estate assets. Survey results from valuers have been analysed to offer strategies in valuation risk management and AI investment decisions in property valuation.

Valuation data and AI evolution

Real property valuation involves different uses. Data types, measurements and sources of data collection can differ. The literature on AI in real estate valuation spans a wide range of topics, from machine learning techniques used in Automated Valuation Models (AVMs) to the integration of AI in investment strategies and market forecasting. The review of the factors in valuing different property types of real property valuation and the key themes that contribute to the academic and practical understanding of AI's role in real estate valuation will help to understand the challenges of AI in real estate property valuation.

Factors in valuing different property types

Traditionally, there are different types of real properties, i.e., residential, commercial, retail and industrial, and specialised properties such as service stations, childcare centres, hotels and hospitals. Due to the shifts in how people work, live and consume and innovations in real estate development and management, some notable emergent real estate asset classes are growing, i.e., data centres, build-to-rent, Airbnb, self-storage, etc. Valuing different types of real estate assets requires consideration of various factors specific to each asset class. As shown in Table 5.1, different data from various sources are required for valuing different real estate asset classes. By carefully evaluating these factors, real estate professionals can arrive at more accurate and comprehensive property valuations tailored to the specific type of assets.

Table 5.1 Factors considered in valuing different types of properties

Real Estate Assets	*Some of the Factors Considered*
Residential	Location, land size, dwelling size, condition, age, number of bedrooms/bathrooms, neighbourhood amenities, school districts and recent comparable sales.
Commercial	Lease terms, rental income potential, tenant quality, location, market demand, zoning regulations, physical attributes, size, NABERS ratings, Green Star, PCA grading and economic indicators.
Retail	Foot traffic, visibility, parking availability, local competition, tenant mix and economic trends affecting consumer spending and income generation potential, Moving Annual Turnover, Catchment analysis, anchor tenants, type of centre, for example, super regional, major regional, neighbourhood, number of levels, degree of competition.
Industrial	Location and accessibility, property size and layout, condition and age of the property, tenant profile and lease structure, income and cash flow potential, functional characteristics and flexibility, market demand and economic trends, environmental and regulatory factors, Property features and amenities, capital and financing availability, future growth potential and development, proximity to infrastructure, road, rail, major transport lines, level of office component to warehouse, clearance, hardstand, loading docks.
Emergent Asset Classes (Data Centre, BTR, etc.)	Market demand and growth potential, location and accessibility, technology integration, regulatory environment, financial metrics and cash flow, risk and market volatility, capital and financing availability, demographic shifts, (Environmental, Social, and Governance) ESG, technological disruption and innovation.

Source: Authors, 2025.

Evolution of AI in property valuation

The evolution of AI in property valuation has progressed through several key phases, influenced by technological advancements, data availability and changes in market practices. Figure 5.1 depicts a chronological overview of this evolution.

Before the 2000s, property valuation was primarily conducted through manual assessments by valuers, relying on comparable sales and lettings, physical inspections and local market knowledge. Limited data analytics tools existed for residential properties, and valuations were often subjective. AVMs were introduced in early 2002, applying algorithms to estimate property values based on comparable sales and property characteristics. While these models were not AI-driven, they marked the beginning of automated processes in valuation. In the mid-2000s, the integration of more sophisticated statistical techniques into AVMs allowed for better accuracy in valuations. Techniques such as regression analysis became common, enabling more complex valuations based on larger datasets.

The rise of machine learning has significantly changed the landscape of property valuation. Between the late 2000s and the early 2010s, algorithms began to incorporate various features, such as economic indicators and demographic data, leading to improved predictive capabilities. This era resulted in the development of more advanced AVMs. The explosion of 'big data' provided access to a wealth of information, including social media, geographic data and economic trends, which emerged in the 2010s. AI technologies began to harness this data, allowing for more nuanced and dynamic property valuations that could account for market fluctuations and consumer behaviour. In the mid-2010s, the introduction of deep learning techniques allowed for even greater complexity in modelling property values. Neural networks could

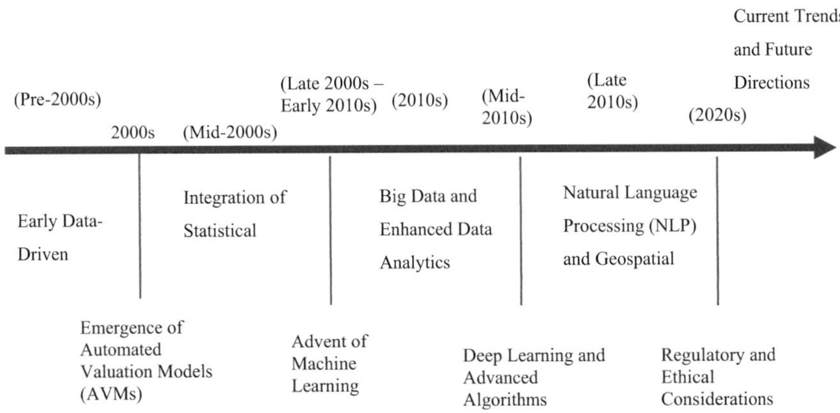

Figure 5.1 Evolution of AI on valuation.

Source: Authors, 2025.

analyse unstructured data (like images and text) alongside traditional structured data, providing a more comprehensive view of property value determinants. In the late 2010s, Natural Language Processing (NLP) technologies enabled the extraction of insights from unstructured data sources, such as property descriptions and reviews. Geospatial analysis tools began to play a crucial role in evaluating location-specific factors affecting property values, enhancing the accuracy of valuations.

In the 2020s, as AI applications in property valuation became more prevalent, concerns around bias (Chiappetta, 2023; Szumilo and Wiegelmann, 2024; Umar and Reuben, 2025), transparency (Cheung, 2024; Abbas and Rinaudo, 2025), and accountability emerged. Industry stakeholders began to emphasise the need for ethical frameworks and regulatory compliance to ensure fair and unbiased valuations. Presently, AI in property valuation continues to evolve with advancements in technology, such as explainable AI, which aims to make AI decision-making more transparent. The use of AI is expanding beyond valuations to include predictive analytics for market trends, risk assessment and investment analysis. Furthermore, ongoing research is focused on enhancing data governance and addressing ethical implications (Toprakli, 2024). As technology continues to advance, AI's role in this field is likely to grow, bringing both opportunities and challenges that regulators, courts and professional bodies must navigate carefully.

Alsahan and AlZaidan (2024) advocated that stakeholders can apply AI to improve efficiency, accuracy and transparency in real estate valuation. However, Stoyanov (2024) suggested that AI could significantly improve the efficiency of the market valuation method, but it should be used as a tool to facilitate analysis, not to replace it altogether. As AI continues to evolve, these contributions provide the foundational knowledge necessary to understand how technology is shaping the future of real estate markets and valuation practices. The next section discusses the research methodology and methods used to understand the challenges and opportunities of data collection using AI in real estate valuation.

Research methodology and methods

This research adopts a qualitative method by studying expert opinions using a purposive non-probability sampling approach (Lucas, 2014). The recognised contributors to the valuation industry experts were selected, for example, well-experienced in valuing properties. We aimed to recruit 12–18 valuation experts to satisfy saturation (Guest et al., 2020). Both close-ended and open-ended questions were answered by the valuation experts. The experts were asked to rank the level of AI used in assisting in the process of valuation, namely, valuation report preparation, data collection, data analysis and data application.

Among the 16 valuation experts who answered the questions anonymously in Qualtrics, most of the valuation experts (around 62.5%) had over 20 years of valuation experience with valuation and land economics backgrounds and hold positions as Valuation Directors, Partners or Senior Valuers in valuation firms. Other valuation experts who are Associate Directors or Valuation Assistants have experience in valuation for between 2 and 16 years. All of them received property valuation-related Bachelor's or Master's qualifications, practice in residential, commercial/retail, industrial/warehouse, development and/or specialised valuations, represent different aspects of the property valuation industry, and offer invaluable insights. One of them has experience in valuing data centres. Table 5.2 depicts the experts' profiles. Thematic analysis (Braun and Clarke, 2006) is used to study the findings.

Findings and discussions about AI in valuation

The findings from the expert responses include the status of AI in valuation firms, the impacts of AI on the valuation industry, AI applications in the valuation industry, ethics and risk concerns of AI in valuation, and the challenges of AI in data collection.

Status of AI in valuation firms

To understand whether valuation firms have used AI in valuation work, two questions were designed:

- Is your organisation already utilising a form of Artificial Intelligence (AI)?
- Are you already using a form of Artificial Intelligence (AI) to undertake valuation work? If yes, what are they? Can you list an example(s) of how Artificial Intelligence (AI) changes valuation processes and reports? If not, does your organisation intend to explore Artificial Intelligence (AI) options?

Twelve answers (75%) confirmed that AI had been used in their firms. Seven (43.7%) valuation experts also confirmed that AI has been used for valuation work. Some firms have built their own AI facilities; others use external software such as Forbury. Expert *L* states that '*we have our in-house AI set up to summarise data and use our internal database of information to produce written work*'. '*Forbury … helps to speed up the valuation process*'. – Expert *K*. '*Pipedrive, a sales customer relationship management tool, is used in one of the firms*' (Expert *C*). Other online websites, such as CoreLogic, Landchecker, Nearmap, Valcre, ChatGPT and LinkedIn, were also mentioned (Expert *C*, Expert *J*). Table 5.3 summarises the portals and software platforms used to undertake valuation work in NSW.

Table 5.2 Valuation experts' profiles

Experts	*Job Title*	*Valuation Experience*	*Valuation Areas*							
			Residential	*Commercial*	*Retail*	*Industrial/ warehouse*	*Development*	*Specialised valuation*	*Data Centre*	*Other*
A	Valuation Director	35	X	X	X	X	X	X		
B	Valuation Partner	30	X	X	X	X	X			X
C	Valuation Director	25	X	X	X	X	X	X		X
D	Valuation Director	31	X	X	X	X	X	X		
E	Senior Valuer	25	X					X		
F	Valuation Director	25		X	X		X	X		
G	Valuation Director	37	X	X	X	X	X			
H	National Director	30	X	X	X	X	X			
I	Senior Manager	40	X	X	X	X	X	X		
J	Assistant Valuer	2		X						
K	Associate Director	7	X			X	X			
L	Valuation Director	12	X				X	X		
M	Associate Director	10		X		X				
N	Associate Director	9		X	X	X	X	X	X	
O	Valuation Director & Advisory	24				X				
P	Certified Property Practitioner	16	X				X			

Source: Authors, 2025.

Table 5.3 Data portals and software platforms used in undertaking valuation work

Data portals and software platforms	*Purpose in valuation work*
Forbury	Used for developing investment valuation models
Landchecker	Used for checking easement, flooding, bushfire, topography, development applications and finding relevant comparable sales
CoreLogic (RP data)	Collecting property sales, transactions and property data information
CoreLogic (Cordell Connect)	Generating proposed development projects in the area
Domain	Collecting property sales, transactions and property data information
Pricefinder	Collecting property sales, transactions and property data information
Nearmap/Matromap	Providing high-resolution aerial maps and location intelligence for calculating houses, roofs, average house size areas, etc.
Valcre	A valuation platform to streamline data gathering (for example, comparable sales, sales and rental analysis), generate quote letters and write reports
PowerBI	An interactive data visualisation software that connects various sources of data (e.g., database, webpage, PDF or structured files like spreadsheets, and SharePoint) into a static and interactive data visualisation dashboard
Pipedrive	Sales customer relationship management tool
LinkedIn	A social media platform for social connection, marketing firms, etc.
ChatGPT	A generative AI chatbot helps write answers

Source: Authors, 2025.

According to Expert F, AI is used 'to assist in reviewing documents and summarising so that we can include the information in the valuation. An example includes Development Application consents and Environmental reports'. Expert C states that AI can be a 'better use of information and analysis to assist the physical inspection'. For the firms that have not yet used AI, Expert G assures that the firm intends to explore AI options but is 'at early stages at the moment'. 'Yes, always willing to explore if the opportunity to use AI fits our business model', Expert C declares. Expert H says that 'for the purposes of analysis of residential projects, we are investigating tools for aggregating, analysing, visualising and sharing data such as Power BI'. 'AI is currently being investigated by our firm with a view of hopefully introducing AI into our daily processes, specifically for valuation', Expert B stated. Expert D states that 'we are currently developing new software with the opportunity to incorporate AI'. Expert A asserts that AI is 'a secondary check method towards quality assurance measures and towards risk management of the business'.

Other software shares special GIS data sets, such as Planning Schemes (contains the policies and provisions that control land use and development and apply to all private and public land), Globe (designs and builds custom software solutions), QImagery (provides an online aerial photograph library for searching and viewing historic images by location) and Contaminated Land Register (a database used to document and track information about potentially contaminated land sites within a particular area).

AI plays an even more critical role in commercial property valuation because these properties are often more complex, higher-stakes and driven by income potential rather than just comparisons. However, the use of AI in the commercial property valuation industry in Australia is still in its infancy. There are no reliable models trained for Australian conditions, as the data is not yet fully discoverable. It would be worth investigating what has happened in the US, as they are usually 5–10 years ahead of other countries. Most of the advances in AI for commercial property are in report writing and presentations of reports. However, it has enormous applications in the areas of report writing, data analysis, quality assurance, preparing for expert evidence and predictive modelling using AI in valuation.

Though some firms have not yet utilised AI, all experts indicated that firms are either developing AI to be used in the near future or are exploring AI applications for their business models, daily valuation processes, quality assurance, visualisation and risk management. This implies that property valuation firms are facing digital transformation pressures, and AI has gradually been playing a more significant role in the valuation industry. One of the biggest problems facing the industry as a result of AI is that education has not caught up with the technology, and as a result, graduates do not have the skills to operate these technologies. Additionally, it will be difficult to retain talent as the industry becomes commodified over time. The impact of AI on the valuation industry will be discussed in the next section.

The impacts of AI on the valuation industry

AI is set to revolutionise real estate valuation by enhancing accuracy, reducing costs and offering deeper insights into market dynamics. Experts pointed out the positive and negative impacts below.

Positive impacts

All participating experts (100%) confirmed that AI will affect the valuation industry. The positive impacts were identified mainly in improving valuation accuracy, efficiency, productivity, cost reduction, data-driven insights and quality assurance.

IMPROVED ACCURACY IN VALUATION

AI leverages vast amounts of data, including historical transaction records, market trends, economic indicators and even neighbourhood sentiment, to provide more precise property valuations. AI models, such as machine learning algorithms, can identify patterns and correlations that may not be immediately obvious to human valuers, leading to more accurate estimates of property values (Abidoye et al., 2021). As Expert *E* states, '*AI produces better quality information and speed*'. '*Through increased efficiency and productivity and a decrease in human error*' (Expert K, Expert L). '*Over a period of time, as databases amass information on all properties in Australia and a process or re-ascertaining the information relating to each property (e.g., every three to five years), the quality of AI valuation models will improve*' (Expert A).

EFFICIENCY AND FASTER VALUATION PROCESSES

> *The efficiency and speed of services will improve. In particular, the mortgage sector of the valuation industry. The improved turnaround in the valuations will lead to financial institutions being able to process and approve loans faster. In the government sector, mass appraisal valuation practices and the reviews of valuation relating to property taxes will improve.*
>
> (Expert A)

Traditional property valuations can take days or even weeks, involving manual assessments, physical inspections and the evaluation of comparable properties. AI can streamline this process by analysing data in real time, drastically reducing the time it takes to generate property valuations. Expert D says that AI '*reduce research time - assists with sales analysis - assists with component development in rating and taxing mass valuation - assists with report writing*'. Expert G advises that AI '*reduce menial tasks of data entry and analysis and allows for time for interpretation of analysis*'. '*It will create better efficiencies, create shorter turnaround times, and enable a valuer to complete jobs faster/quicker*' (Expert O).

Most of the experts have mentioned that AI will improve efficiency in terms of '*producing basic reports*', '*simplifying mundane tasks*' (Expert M), '*improving turn-around times*' (Expert I), '*speed up the data process*' (Expert P), '*the review and analysis of data*' (Expert F) and '*repetitive tasks such as Q&A processes*' (Expert B). Automated Valuation Models (AVMs) can now provide instant estimates based on real-time data.

DATA-DRIVEN INSIGHTS

'*Increase productivity dealing with data, which is paramount to the Valuation industry*' (Expert C). '*Valuers can retrieve more information more easily*' (Expert J). AI helps real estate professionals gain deeper insights into factors that affect property values (Alsahan and AlZaidan, 2024). For instance, AI can analyse demographic trends, local economic conditions, neighbourhood developments and even social media activity to predict how these factors might influence future property values. This is valuable for investors, developers and lenders who need to assess potential returns on investment or risk.

QUALITY ASSURANCE AND CUSTOMISATION OF VALUATIONS

Expert A states that

> *AI models integrated within valuation management platforms will enhance quality assurance measures, reducing risk, leading possibly to a reduced professional indemnity insurance cost to a business. Industry Professional Standards will likely change within the valuation sector. The professional standards organisation will integrate AI to accommodate the new age valuation practices that will develop.*

AI models can be tailored to consider specific features of properties that are highly relevant to certain buyers or investors, such as proximity to schools, transportation hubs or other amenities. This allows for more personalised and precise property valuations, especially in diverse markets where one-size-fits-all valuation models may not be as effective.

COST EFFICIENCY

> *The larger firms operating in the volume-based mortgage valuation oligopoly market will benefit from a reduced cost base, leading to increased profits. AI will assist in the reduction of their professional employment costs (Valuers), leading to a possible increase in profits from the current labour-intensive process.*
>
> (Expert A)

AVMs powered by AI can significantly reduce costs for real estate firms and valuers. Traditional methods of property valuation involve multiple steps and a team of professionals, whereas AI systems can provide similar results at a fraction of the cost and with fewer resources. Experts H and J state, '*Time and cost savings to the business, speed to market*'.

In summary, the identified positive impacts of AI in valuation, i.e., improving data accuracy and efficiency, are aligned with the literature (Alsawan and Alshurideh, 2022; Mathotaarachchi et al., 2024). The experts have emphasised that AVMs help to improve valuation efficiency through the fast processing of data. This brings important data collection and quality issues.

Negative impacts

The integration of Artificial Intelligence (AI) into real estate valuation brings potential drawbacks, including job loss, overreliance on data, lack of human intuition, reduction in competition, potential fraud and biases and valuation standard and ethics issues. Also, it will discourage new graduates from joining the valuation industry, thus leading to brain drain.

LOSS OF JOBS AND EXPERT INPUT

The role of the valuer will change from conducting analysis in the comparison process and applying appropriate valuation techniques in the residential sector to becoming a property data collector (Expert A). The rise of AI in real estate valuation could lead to job losses, especially for traditional roles like valuers, analysts, or real estate agents (Lee et al., 2024), while opening up new jobs in Proptech. AI can automate much of the valuation process, reducing the need for human input and '*replacement of skilled professionals by machine learning techniques*' (Expert H). '*It may, in some instances of repetitive work, take away some roles*' (Expert F).

AI-driven processes may make real estate transactions feel less personal, as buyers and sellers rely more on machines than human professionals for key decisions. This could decrease trust in the process and alienate some users. '*Potential valuer loss, lack of personal input into valuations and not sure, but most likely high cost to businesses initially implementing AI*' (Expert B). Potential bias could be an issue. '*Some loss of employment and also loss of valuation skills over time*' (Expert C).

OVERRELIANCE ON DATA AND ALGORITHMS

AI systems depend on large datasets to make predictions. If these datasets are incomplete, outdated, or biased (e.g., under-representing certain neighbourhoods or demographics), the AI's valuation could be skewed, leading to inaccurate or unfair property assessments. AI models may overlook certain qualitative factors that affect property values, such as local community dynamics, neighbourhood reputation or future developments that are not easily quantifiable. As indicated by Expert *J, 'AI might still produce the wrong information and information can be easily hacked, so valuers need to proceed with caution'*. When a valuation is in disagreement, a property owner can provide their evidence to a valuer for review. How can the property owner challenge a valuation assessment to AI if the report is generated by a machine?

In Australia, CoreLogic estimates residential property value by range based on their available datasets, onthehouse.com and propertyvalue.com.au. Another AI property value prediction platform is Domain, domain.com.au, which provides low, median and high estimations for residential properties. Domain also owns PriceFinder, pricefinder.com.au, which provides information to real estate professionals and enterprises. The authors collected data randomly and compared the estimated prices from CoreLogic and Domain estimations before the property auctions with their actual sold prices after the auctions.

Table 5.4 shows the 12 property auctions between 22 February and 27 February 2025. Some estimations could not be found at the time of the search on propertyvalue.com. It is found that the AI-estimated property values are not the same between the two websites. The actual sold prices were relatively close but higher or lower than the predicted values. This implies that AI's valuation could be biased, as some qualitative factors might not have been considered in the AI estimations. To ensure reliable and valid property assessments when integrating multiple sources with conflicting data, for example, zoning discrepancies, real-time data reconciliation measures should be implemented, including immediate identification of discrepancies, data validation through third-party references, continuous updates from multiple sources, and set priority rules to determine which data source should take precedence in case of conflicts, i.e., official zoning information from local government records may override other sources.

LACK OF HUMAN INTUITION

AI can also overlook human factors. AI is built on historical data and patterns, but may be trained to capture human intuition or emotional factors that influence real estate markets. Things like local politics, unforeseen infrastructure changes or changes in consumer preferences are hard to predict accurately by an AI system. Personal needs can be disregarded. Real estate valuation is often tied to personal circumstances, and AI might miss personal preferences or unique property features that aren't quantifiable in datasets but are important to buyers and sellers.

> *I doubt whether AI could write full report sections because of the unique nature of every single report. At least this would be so for commercial valuations in my view. …perhaps AI could write the basic and somewhat repetitive report sections in commercial valuations…*
>
> (Expert *E*)

REDUCED COMPETITION AND CONSUMER CHOICE

> *AI will likely lead to further aggregation of the sector - leading to reduced competition and consumer choice. Smaller firms are folding as their practices are not saleable due to the strong goodwill attached to the practice partners. Further, the capital cost to introduce AI into the business outweighs the benefits that relate to sustainability, again leading to reduced competition and consumer choice of experienced sector-based boutique valuation firms. A potential exists that valuation firms will cease to operate in their current form and be replaced by property data companies.*
>
> (Expert A)

Table 5.4 Quantitative analysis of AI valuations and market-sold evidence

	Property address	*Property feature*	*Property value / on the house*	*Domain ($)*	*Actual Sold price*
1	14 Heber Street, Hurstville, NSW 2220	5 beds, 3 baths, 2 cars, 607 sqm/597 sqm	$2.6m–$2.8m	2.39m–2.75m–3.11m	$2,805,000
2	31 Evelyn Street North, Sylvania NSW 2224	5 beds, 3 baths, 5 cars, 448 sqm		2.97m–3.39m–3.81m	$3,240,000
3	28 & 28A Becharry Road, Blacktown NSW 2148	4 beds, 2 baths, 1 car, 556 sqm	$1.089m	910k–1.06m–1.21m	$1,206,000
4	6 Wellesley Crescent, Kings Park, NSW 2148	4 beds, 1 bath, 1 car, 573 sqm	$1m–$1.1m	1.05m–1.22m–1.39m	$1,080,000
5	3/83 Foster st, Leichhardt, NSW 2040	3 beds, 2 baths, 1 bed, 149 sqm		2.2m–2.54m–2.88m	$1,435,001
6	112 Hoof st, Grafton, NSW 2460	3 beds, 1 bath, 1 car, 588 sqm	$325k–$350k	380k–440k–500k	$365,000
7	7 Newbolt st, Wetherill Park, NSW 2164	4 beds, 2 baths, 2 cars, 680.9 sqm		1.52m–1.76m–2m	$1,751,000
8	26 Shawnee St, Greenfield Park, NSW 2176	3 beds, 1 bath, 1 car, 574.4 sqm		1.1m–1.28m–1.46m	$1,300,000
9	24 Welwyn Road, Canley Heights, NSW 2166	3 beds, 1 bath, 2 cars, 562.8 sqm		1.02m–1.18m–1.34m	$1,065,000
10	6b Earl Street, Hunters Hill, NSW 2110	4 beds, 2 baths, 2 cars, 334 sqm	$4.192m/$4.75m	3.37m–4.42m–4.47m	$4,920,000
11	7 Barripa Close, North Ryde, NSW 2113	5 beds, 4 baths, 2 cars, 290 sqm	$2.601	2.04m–2.36m–2.68m	$2.300,000
12	24/34–38 Gerard st, Cremorne, NSW 2090	1 bed, 1 bath, 1 car apartment	$733k	650k–760k–870k	$680k offer

Source: Authors, 2025.

HOMOGENISED VALUATION AND INFLEXIBILITY

AI tends to create standardised models that may overlook the unique attributes of properties that aren't captured in historical data. The model may give equal weight to properties in different locations despite clear differences in desirability or amenities.

> *For AI to have high levels of quality in the market, it will require more detailed analysis of sales evidence attributes entered into databases based upon physical verification of property by what is known as a valuer in geographic locations, where more intrinsic data is collected (than currently captured in Valuation platforms) can be compared to a property which is valued when the attribute of a subject property is valued.*
>
> (Expert A)

AI-based systems may struggle to consider all the nuanced factors that a human valuer might, leading to a less nuanced valuation that doesn't account for 'soft' factors like charm, location dynamics or future development prospects.

ETHICAL ISSUES IN AUTOMATION

As AI-driven valuations replace human valuers and other professionals, questions arise about the fairness of these changes.

> *Currently, there are several Property Data collecting companies (e.g., RP Data, Domain/Pricefinder etc), which have the ability to create valuation models, therefore the potential to augment their data collection process and overlay information from various open source government agency planning, spatial information along with the creation of a property valuer /surveyor service to enhance intrinsic attributes of property data.... which may possibly lead to the collapse of the "Valuation" industry current practice structure.*
>
> (Expert A)

This is likely in the not-too-distant future.

VALUATION STANDARDS

The Royal Institution of Chartered Surveyors (RICS) has updated the use of AI in valuation in the Global Red Book, which has been effective since 31 January 2025. The AI regulation includes a commitment to transparency in its use, consistent professional judgement in its application, and confidentiality and data rights protections (RICS, 2024). However, the policy documents on the adoption of AI in valuations have not been found in the Australian

Property Institute (API). '*If AI is being pushed in the valuation industry, the same industry would have to review its current standards. For example, the required onsite inspection a valuer must complete in person could not be completed by AI*' (Expert *C*).

> *Valuation practice has developed through professional standards and legal precedents that have been derived through various levels of court in determining market value. The adoption of AI-based valuation methodologies relating to property taxes may likely impact the process of a landowner appealing values issued by the Government, where the current lawful process requires the physical analysis of sales evidence and adjustments of these sales to compare to a property by a valuer.*
>
> (Expert A)

Though there are many advantages of AI in valuation, the API and the International Valuation Standards Council (IVSC) have not caught up with machine learning in real estate valuation for commercial properties, i.e., it is not an accepted method of valuation. Apart from valuation standards, homogenised reports, etc., the potential for data bias leading to discriminatory valuations (Topraklı, 2024) of AI in valuation was the main concern of the experts. In addition, liability is another negative factor when AI produces an incorrect valuation output that causes a loss to consumers. How does AI sit within the requirements of a valuation firm's professional indemnity insurance, which has not been addressed in the current valuation practice? This negative impact must be addressed in the transformation of AI in the valuation industry.

Replacement of AI in the valuation industry and thought process

Two different views on whether AI can replace the valuation industry and the thought process of a valuer to estimate a value independently of human involvement. Expert *F* states

> *The valuation industry as a collective may decide that AI valuations are not acceptable, even if the capacity is available. Think accountability. Could you hold AI accountable for under- or over-valuing a property in the same way, valuers can currently be held accountable for their work? And what about the (still valid) requirement for valuers to prepare a 'speaking valuation'? Could AI do this? Perhaps not, as there is no insight into the AI's reasoning. It's a black box. Overall, I doubt whether the industry would let AI take over too much (even if the capacity is there). Again, perhaps in strongly homogenous residential sectors but less likely in non-residential valuations.*

The points made by the experts are aligned with Stoyanov (2024), indicating AI should be used as a tool to facilitate analysis but not to replace it altogether at this point in time. A different view is that if the systems are reliable and create the same results as a human valuer, then the consumers will accept the product. The question of whether AI can be held liable is not the issue, as the makers of the AI model could have professional indemnity insurance, as a human valuer does.

Two of 16 experts support that AI can replace the thought of a valuer to estimate a property value. The argument is that

> *in instances where a valuation is not required for legal or tax purposes, i.e. the mortgage market (which makes up the highest proportion of services valuers provide in Australia). An AI process will require extensive IT capital investment to create valuation methodology templates that relate to each asset category, along with specialised and more intrinsic data collection for "field" valuers or qualified data collectors to capture property attributes, interconnecting with government agency planning and government agency spatial overlays. Statistical QA process integrated with IT software would enhance the reliability of the result for the consumer. Such a process can be achieved at different levels relative to the asset type, e.g., residential, commercial, rural specialised property).*
>
> (Expert A)

'*In some circumstances, yes, but at this time, it would be limited to homogeneous products (e.g., housing stock) and non-specialised commercial property*' (Expert H).

Most experts are against the statement. The reason is that

> *the capability may be there to "spit out" a number, but I refer to the above point of accountability. And also trust and transparency, and let's not forget client service. And what about adopting the correct valuation approach for the subject property? There is no matrix for that, as it is always asset-specific. Clients often want to know why the adopted yield is X and not Y. Could AI explain its own reasoning in human terms? Can one be certain that all traditional aspects of valuation have been considered by AI? There are many, and many are "subjective", such as the perceived quality of the tenant.*
>
> (Expert F)

An argument is that '*AI is very good at explaining details*'.

Most experts recognise that AI cannot replace human valuation in the short term, apart from valuing homogeneous products like mortgage valuations, due to the 'black box' of processing the data by AI. The concerns were built on fairness, transparency and accountability in valuation services. AVMs can replace human valuation when data collection captures property attributes, and data quality can be ensured in the long term.

Residential automated valuation models (AVMs) for commercial valuation

Four of 16 experts support that AVMs in the residential sphere are suitable for use in the commercial or specialised spheres '*when a low-risk valuation is involved, for mortgage work if lending ratios are below 40%*' (Expert *G*). AVMs can be used in low-cost commercial property under certain value ranges, such as shops or small industrial strata units. *This would require robust data collection of market evidence along with interconnecting the attributes of the property being valued to the asset market data to create a reliable scorecard and identify outliers* (Expert *A*).

> *I think AI can assist with commercial valuations, however, the institutional property sector is highly specialised and sophisticated, where one size does NOT fit all and will always take human input to derive appropriate metrics for an asset. Hence, although AI can help, it will not take over this sector of valuations.*
>
> (Expert *O*)

'*There is a place for it to improve efficiencies and margin for error, however, it won't completely take over given the objective nature of much of the work*' (Expert L).

Most experts are not confident in applying the residential AVMs for commercial valuation due to concerns about valuation accuracy.

> *AI valuations for low-value residential (i.e. apartments) are viable. AI valuations for commercial properties are highly problematic, as the data sets relied upon do not have the depth that residential data sets have. i.e. when valuing land, there may only be one or two transactions.*
>
> (Expert N)

'*It would need to become more reliable and accurate than it is currently to be used on assets with a significantly higher value than residential houses*' (Expert K). '*There are too many considerations with too few datasets to rely on in the specialised commercial space*' (Expert N). '*... In commercial or specialised fields, there would be limited use due to the lack of sales evidence and property variability*' (Expert D). '*There are too many moving components and opinion-based calls that need to be made by a human*' (Expert M).

> *... Some tenants pay a net rent, a gross rent or increases in outgoings which require the valuer to read all lease documents for all tenancies. Ascribing a market rent also needs a level of judgment to compare properties on a like-for-like basis. Outgoings also need to be analysed and compared to industry benchmarks.*
>
> (Expert I)

The second reason is that '… AVMs are not valuations but rather assessments and therefore should not be part of any commercial valuation' (Expert E).

> From my knowledge, the accuracy of these models has seen minimal improvement over more recent times. Accuracy is a concern in areas/localities with high variability in property attributes and limited availability of sales evidence. … they are not considered appropriate where a high level of accuracy is required other than for Quality Assurance purposes.
>
> (Expert D)

> Notwithstanding an automated result in value, …AI modelling result would require the intervention of a valuer, where sanitation and analysis of the salient characteristics of the subject property can be conducted on a subject property, which would include, but not be limited to, the tenancy of the subject property. Such a process would require a level of procedural QA whereby a valuer must have geographic knowledge and a level of experience with the asset being valued.
>
> (Expert A)

This point may be valid in Australia, but it may be worth investigating the American experience. Figure 5.2 Word Cloud summarises the points made by the experts.

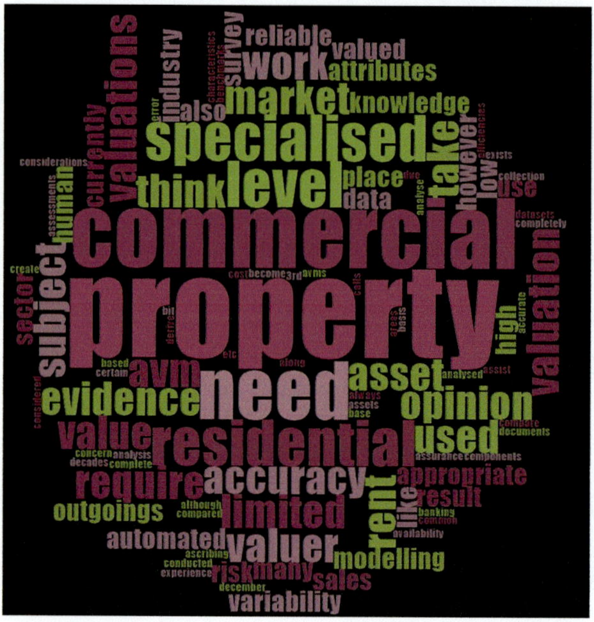

Figure 5.2 Residential AVMs for commercial valuation.

Source: Authors, 2025.

Incorporating AVMs into property valuations can be highly beneficial for many purposes, including speeding up the process compared to traditional methods, cost-effectiveness, consistency and standardisation and scalability to handle large volumes of property data and generate valuations for numerous properties simultaneously. Despite these benefits, AVMs have limitations that prevent them from being fully accepted in formal lending practices, particularly by Australian banks. The key limitations include a) a lack of physical inspection as AVMs typically rely on existing data such as sales history, location, size and other publicly available information; b) a lack of quality data and some commercial data are not available, such as commercial sales evidence, lease terms, incentives, etc.; c) AVMs are automated and do not involve human judgement and thus cannot meet the requirements for formal valuations needed to secure lending; d) AVMs can lead to outdated or inaccurate valuations as AVMs may not be quick enough to reflect market volatility and changes of local factors and e) a risk of misvaluing properties due to incorrect assumptions or modelling errors. Given the limitations of AVMs, which are currently used for low-level lending in the residential sector, AVMs could be used for low-cost base commercial properties under certain value ranges, such as shop tops or industrial strata units, where property features and transaction data are easily collected. AVMs can also be used as a supplemental tool for initial property assessments, risk profiling and portfolio valuation.

Challenges and risks of AI in the valuation industry

Commercial real estate is still lagging behind residential in terms of data availability and accessibility. All the experts agree that AI introduces certain risks that need to be carefully managed. Though there are definitely additional risks associated with AI, this is unlikely to stop the use of AI in valuation. Some key risks that AI can bring to the valuation industry include.

Bias and discrimination

AI systems learn from historical data. If the data used to train AI models is biased (e.g., biased property prices, past valuation errors, or demographic imbalances), the AI may replicate and amplify those biases (McCord, 2022). Even if the training data is unbiased, the algorithms used in AI models can introduce their own biases based on how they are structured or how features are weighted. These biases could result in valuations that disproportionately affect certain groups or types of assets. '*Incorrect use of data may provide a poor result*' (Expert *F*).

Lack of transparency and explainability

Many AI models, especially deep learning models, are highly complex and function like 'black boxes' where it is difficult to understand how decisions or valuations are being made. This lack of transparency (Cheung, 2024) can undermine trust in AI-generated valuations and make it difficult to identify the reasons behind specific valuation results. In the valuation industry, particularly in areas like real estate or financial services, stakeholders may require clear explanations for valuation outcomes (e.g., lenders, investors, clients). If an AI model can't provide understandable explanations for how it arrived at a particular valuation, it could lead to disputes, reduced trust or legal challenges. The risk of AI on valuation is an '*error in the output/AI assumptions*' (Expert M).

Overreliance on data

Valuation results '*can be biased if the data doesn't support it*' (Expert P). AI's effectiveness depends on the quality and accuracy of the data it processes. If the data is incomplete, outdated or inaccurate, the AI model will produce inaccurate valuations. In real estate, for example, if market trends are not properly captured or data sources are flawed, the AI could generate misleading property values. If an AI model is too finely tuned to historical data, it may become overfitted, meaning it works well on past data but struggles to adapt to new, unseen data. This can lead to poor performance, especially in volatile or dynamic markets.

> *Perception is often reality. If AI models are utilised on a broad scale, especially if used by parties that are uneducated, then this could artificially skew the market. – People who only rely on these models could make decisions which have a negative financial impact – Loss of valuers who currently play a vital role in managing risk and providing advice for their clients.*
>
> (Expert D)

> *Incorrect use of data may provide a poor result.*
>
> (Expert F)

Fraud and security risks

AI models are vulnerable to data manipulation, whether intentional (e.g., fraudulent actors providing false data inputs) or accidental. If an AI model relies on manipulated data to make its valuation predictions, it can be tricked into providing inaccurate or fraudulent results. For example, a failed foundation, flood damage or the property next door, owned and occupied by a heavy metal rock band, may not be included in a sale contract. Without providing the information, a AI or AVM valuation will not reflect those negative factors in the valuation process. To prevent false data input, regulatory compliance should be requested to conduct data source checking and validation in the AI system. At the same time, human valuers should also play a role in validating the data correctness and providing valuation judgement.

AI systems are often connected to large datasets and cloud-based platforms, making them potential targets for cyberattacks. If an AI system is hacked, it could result in unauthorised changes to valuations, leading to financial losses or legal issues. *'Cybersecurity risks, lack of creativity, especially regarding risk concerns within valuations, lack of transparency & privacy concerns'* (Expert H). Privacy laws and confidentiality provisions play a significant role in shaping market transparency in Australia, particularly in the context of commercial transactions, real estate incentives and valuation accuracy. These regulations are vital in protecting individual and business information, but they also present challenges when it comes to achieving full transparency in the property market.

In Australia, privacy is governed by the Privacy Act 1988 (Cth) and various state-based privacy laws. These laws focus on protecting personal information, particularly that which can identify individuals or businesses. The confidentiality provisions are common in real estate transactions, especially in commercial deals. These provisions are usually embedded in agreements to ensure that sensitive business or financial information remains confidential between the parties involved. For example, real estate incentives—such as rent incentives, rebates and subsidies—are often employed to stimulate investment and support market growth. The incentive can skew the market and lead to valuation inaccuracies if not disclosed. Lack of transparency about the role of incentives can lead to overinflated valuations, or the risk profile of a property might be miscalculated, leading to errors in the valuation process. Thus, striking a balance between protecting sensitive information and ensuring enough data is disclosed for accurate valuation and market valuation is key.

Loss of human expertise and judgement

A material risk of AI in valuation is '*Replacing valuation jobs*' (Expert K). As AI takes over more of the valuation processes, there's a risk of undervaluing human expertise, particularly in situations where subjective judgement is required. AI may struggle with factors like local market knowledge, cultural context or nuanced characteristics of a property or asset that human valuers would consider. Overreliance on AI could result in valuations that lack the insight that human experts bring to the table.

AI's growing role in valuation processes may also result in the displacement of human professionals, particularly valuers, analysts and other roles that involve judgement-based decision-making. While AI can improve efficiency, the automation of valuation work could disrupt traditional jobs, creating economic and workforce challenges. While AI can enhance efficiency, it's essential to balance its use with the insights of experienced valuers who understand local markets and nuances (Stang, 2023).

Inadequate regulation and ethical concerns

The valuation industry, especially in real estate and finance, may lack the regulatory frameworks needed to ensure that AI models are being used ethically and consistently. The use of AI in property valuation raises ethical questions about privacy, data security and the implications of automating decisions that significantly impact people's lives, such as mortgage approvals or property investments (Amorin, 2024; Yiu and Cheung, 2024). The majority of location-based intelligence has evolved substantially over the last 5 years and now includes mass mobility data (e.g., captured from open apps in mobile phones), credit card data and facial recognition data, all of which create serious ethical concerns. However, these tools provide far more analytics and data reliability than a valuer can and are likely to be significantly more relied upon over time. Without industry standards and regulations, AI-generated valuations could vary widely, leading to confusion, disputes and potentially unfair practices.

AI models could be used in ways that exploit vulnerable individuals, such as giving certain property buyers inflated valuations or encouraging risky financial decisions. If AI is deployed without considering ethical standards, it could perpetuate or exacerbate inequalities in the valuation process. '*Data breach, removal of intuition/opinion that is critical in valuations, limited education of new valuers coming in*' (Expert L). '*Fraud, Inaccurate or incorrect information, frustration on the client side*' (Expert J).

Overconfidence in AI outputs

As AI-generated valuations are perceived as objective and data-driven, there may be a tendency to trust them without question. Overconfidence in AI outputs could lead to poor decision-making, especially if AI systems are not fully understood by the users or are relied upon too heavily without proper oversight. '*Data breaches and inability to determine key inputs of a commercial valuation, which requires human input, social aspects, discussions, case-by-case valuation situations, inaccuracies in modelling and reporting*' (Expert O).

AI can produce results that look accurate on the surface but are actually flawed. For example, an AI system may generate a valuation that fits historical patterns but misses emerging market trends, leading to false confidence in its outputs. This is particularly dangerous if stakeholders rely on these outputs for high-stakes decisions (e.g., real estate investments, insurance underwriting). '*If the data relied upon in the AI valuation report is not properly analysed, verified, inspected and compared to the subject property, on a like-for-like basis, then the potential for error is highly likely*' (Expert I).

MARKET VOLATILITY AND ADAPTATION

AI models trained on historical data may struggle to adapt to sudden shifts in market conditions, such as an economic downturn, regulatory changes, or unexpected events (e.g., natural disasters, pandemics). AI models may struggle to adapt to sudden changes, such as economic downturns or shifts in consumer preferences (Behera et al., 2023), and the recent COVID-19 pandemic had an immediate impact on the property market, where tenants were unable to continue operating. If many companies or professionals in the industry adopt similar AI systems, there's a risk that the entire market could become overly reliant on similar data sources and algorithms, leading to a form of 'herding' behaviour. This could result in mass misvaluations or market bubbles, as AI systems may all respond similarly to the same signals. It is possible that different AI systems will produce different results.

> *It can reduce the requirement for actual valuers. Like modelling, it can produce very erratic answers. The fewer valuers are in the field analysing sales, etc., the less valuers may have an understanding of the market. Valuers currently speak with owners, purchasers, developers, real estate agents, lenders, etc., to gain a better understanding of market drivers and value.*
>
> (Expert G)

LEGAL AND LIABILITY RISKS

If an AI system produces an inaccurate valuation that leads to financial loss, it's unclear who would be held liable. Is it the responsibility of the company that deployed the AI, the developers who created the system, or the AI itself? In many jurisdictions, there is still uncertainty regarding the legal responsibility for AI-generated outputs. Will professional indemnity insurers cover AI-generated valuations?

> *In areas of valuation where legal or statutory challenges are made in relation to a property the use of AI may not be challenged and in the longer term as the valuer role changes, there may not be many 'expert' valuers who can analyse value with the legal precedents that apply to determine a market value (in accordance with that legal definition).*
>
> (Expert A)

Valuation industries like real estate, insurance and finance are often highly regulated. If AI models are not properly aligned with local regulations or fail to account for legal nuances, they could lead to compliance issues or legal challenges. '*The review process of AI-generated reports may not be as stringent. AI may not cut through fraud etc., dodgy sale contracts, for example*' (Expert C).

In summary, while AI has the potential to bring significant benefits to the valuation industry, these risks highlight the importance of carefully managing the integration of AI. Addressing these challenges requires a balanced approach, including ensuring valid data collection, improving transparency, ensuring regulatory oversight, maintaining human expertise and prioritising fairness and ethics. With proper safeguards, AI can help streamline the valuation process and improve accuracy, but it must be used thoughtfully to avoid unintended consequences.

AI contributions to valuation stages

Real estate valuation relies on collecting relevant data, for example, evidence of rent or sales, property specifics, micro- and macroeconomic factors, and so on; analysing the collected data, the valuation report and the valuation decision, for example, applying yield, rate per square metre, etc., to determine property values. To understand how AI contributes to each stage of valuation, the experts ranked the importance of AI in the valuation stages, where '1' indicates less assistance and '5' means significant assistance.

As shown in Figure 5.3, 61% of the experts strongly support that AI can be utilised in the valuation preparation stage to search property information like planning controls and property physical attributes, for example, a map that shows the land location. Around one-third of experts support AI's ability to assist with data collection. An expert strongly rejects the AI's role in collecting valuation data. This is because AI may source publicly available data, for example, suburb profiles or property sales or lease information from various sources; however, some business data, such as rental income generated by existing tenants, incentives, WALEs, occupancy rates, operating expenses, etc., may be difficult to collect. Missing such information may provide biased valuation results.

In summary, though AI may provide analytical tools, most experts affirm that AI has fewer contributions to data analysis, especially in making valuation decisions. As discussed previously, AI may be able to provide an efficient way to analyse data and generate a value based on the provided data; however, AI may not be able to determine what variables should be emphasised in valuing different types of properties. Human experts and judgement are still playing an important role in valuation decision-making.

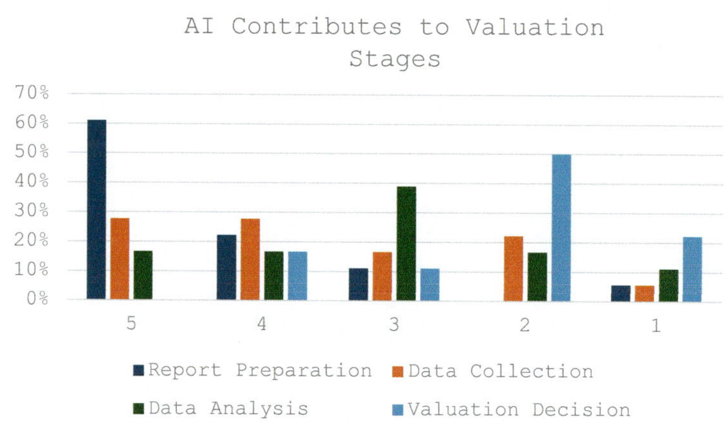

Figure 5.3 AI impact of data in valuation (Source: Authors, 2025).

Conclusion, recommendation and further research

This research paper has studied AI's role in real estate valuation through a survey of valuation experts' views. Some conclusions have been generated by investigating the impact of AI on real estate valuation, whether AI can replace the property valuation industry and human involvement, whether AI in residential valuation can apply in commercial spheres, whether AI is susceptible to fraud and bias, whether AI valuation can achieve the same level of human-generated accuracy and the risk of AI in valuation decision-making.

Conclusion 1: AI plays an important role in modern valuation practices. Improvement of data collection and processing, and reduction in business costs were positive impacts of AI on real estate valuation identified. Loss of human judgement, data bias and potential data manipulation, job replacement, privacy and ethics concerns were the negative impacts. This implies that while AI can bring efficiency and precision to real estate valuation, the negative impacts should be noted.

Conclusion 2: AI may continue to evolve, but the future of property valuation remains uncertain. However, AI can be trained but cannot replace the real estate valuation industry completely, and human-generated valuation will always be essential. While AI may offer efficiency, it cannot fully replicate the depth of analysis and personalised service provided by human property valuers in the short term. However, it is the trend that there will be a need for fewer valuers, and those that are needed will do highly specialised work in the future.

Conclusion 3: The use of AI in real estate valuation can introduce challenges related to data privacy, bias and transparency. For instance, if the data used to train AI models is biased or incomplete, this can affect the fairness of valuations, particularly in marginalised neighbourhoods. Regulators may need to create frameworks to ensure that AI models are fair and unbiased in their predictions.

Conclusion 4: Human-generated valuations may still hold higher confidence levels due to their perceived reliability and the human capacity for judgement in the short term. Ultimately, the level of confidence will vary depending on the industry, the specific use cases and the level of understanding and trust in AI systems. However, AI-generated valuations can gain equal confidence over time if they demonstrate consistent accuracy and transparency and are used in appropriate contexts.

Conclusion 5: While AI-generated valuations can provide efficiency and consistency, they are not immune to fraud. If the systems are poorly designed, inadequately supervised or exploited by malicious actors, they could be used to carry out fraudulent activities. However, with proper safeguards, such as data verification, regulatory oversight and human auditing, the risks of fraud can be significantly reduced.

The findings of this research paper provide insights into the challenges and opportunities of data collection in the valuation industry that have contributed to the digital transformation of emerging AI technologies in property and asset valuation. In theory, AI technology can process large quantities of property transaction data and value properties automatically. In practice, using AI in the valuation industry is in its infancy and a transitional stage. Data availability and quality are critical in the valuation. Due to privacy issues, some building owners do not want to release information to the public, for example, rent incentives. There is also no checking system for sources of data and no validation process built into AI systems, thus far. Some valuers are reluctant to apply AI in conducting valuations, especially for valuing commercial properties in the short term. However, in the longer term, with technological advancement, valuation automation is a trend that will be gradually applied to the valuation industry. The traditional valuation report may be replaced by a new format that includes much more detail and useful data and could cater to multiple scenarios. In addition, the younger valuers may have learned the new technology faster, but some older valuers may take time to adopt AI in valuation practice or resist it completely. Bringing the contents of AI in the real estate industry into teaching and learning in education will lead to a broader understanding of how AI is used positively in the valuation practice.

Several implications and recommendations can be derived from this research. First, AI is transforming the valuation profession, offering increased accuracy, efficiency and predictive power, as well as opening up new opportunities for valuers to offer advanced services such as predictive analytics, real-time market monitoring and dynamic asset valuation models that can be tailored to specific client needs. However, it also presents challenges related to transparency, trust, data ethics and skill requirements. Valuers should understand the roles AI plays in valuation and its limitations. The valuers should learn how AI can assist with valuation and engage with data analysis and valuation decision-making to improve the accuracy and efficiency of valuation. Valuers should also understand the need for their judgement and expertise, especially in situations where nuanced human insights are required to interpret complex market conditions or value non-standard assets. At the same time, valuers should be informed of any changes and updates on industry standards and guidelines for valuations due to AI tools being integrated into valuation practices. Professional organisations such as the Australian Property Institute (API) and the Royal Institution of Chartered Surveyors (RICS) may introduce new certification programmes or qualifications for valuers to demonstrate their competency in working with AI systems.

Secondly, AI developers play critical roles in designing and selecting the appropriate machine learning algorithms that will drive the AI's valuation models. The traditional valuation report may be replaced by a new format that includes much more detail and useful data and could cater to multiple scenarios in the future. Valuation reports have hardly changed in over a century;

therefore, they will likely need to evolve. The AI developers should understand the challenges of using AI to collect real estate data to ensure the data collected is accurate, complete and reliable. Many real estate data sources, for example, public records or proprietary databases, may have restrictions on access. Some data might not be available digitally or could be fragmented across multiple platforms. Real estate data also stems from various sources, including public records, MLS listings, social media and economic reports. Integrating these diverse datasets into a cohesive format suitable for AI analysis can be challenging. Dynamic market conditions, data bias produced from AI models, and ethical considerations should be built into the implementation of AI in property valuation's multifaceted approach. The AI developers should also implement robust security protocols to protect sensitive data used in AI valuation models. By implementing a role-based access control system, only authorised individuals can modify the AI models or access sensitive valuation data.

Thirdly, property organisations, including valuation firms, need to recognise the unstoppable AI generation in the valuation process by combining robust technical solutions, comprehensive policies (e.g., ethics standards, data governance framework, etc.), and ongoing monitoring measures (e.g., ethics audit, performance matrix, etc.) to effectively leverage AI in property valuation while minimising risks and enhancing the accuracy and fairness of valuation practice. It is recommended that commercial valuation firms pragmatically adopt AI systems by determining where AI could add value to the firm's existing processes and focusing on specific areas where AI can provide clear, measurable benefits and scale up gradually as the firm gains experience and confidence in the technology. By working with AI vendors, valuation firms can select and implement AI solutions tailored to their needs. Testing AI-driven valuations against traditional methods and fine-tuning the system based on feedback. Valuation firms should also develop a robust data governance framework and data auditing system to ensure data accuracy, relevance and compliance. While AI handles large datasets and complex valuations, human expertise should remain an essential part of the decision-making process. In addition, valuation firms should establish guidelines on the ethical use of AI and ensure that AI systems used in valuations are transparent. The valuation firms should also ensure compliance with regulations and keep detailed records of AI models, data sources, and decision-making processes and regularly audit to demonstrate compliance.

Fourthly, it is the industry and government's responsibility to set up regulatory frameworks to ensure AI applications in valuation are ethically transparent and beneficial to all stakeholders, including protecting consumers, ensuring accuracy, fostering innovation and addressing the impact of automation on the workforce. AI systems used in valuation often rely on enormous datasets. Governments, both State and Federal, hold large amounts of data that could be made available to train valuation models. For example, in the aged care sector, there is an enormous amount of pricing data available, yet the Federal Government refuses to allow the data to be used for commercial purposes. It is recommended that governments have a role in ensuring that the data used is ethically sourced, representative and compliant with privacy laws. The government can also foster innovation in AI by supporting research valuation practices. For example, blockchain technology that stores data in blocks and allows secure and transparent information sharing among stakeholders can be proposed for managing property transactions, data accuracy and ethical practice and mitigating risks in valuation.

AI application in the property valuation industry is in its infancy stage, currently being used in residential and mortgage valuation, but yet to be adopted in commercial property valuation. Commercial properties are often valued based on income generation (e.g., cap rate, NOI). AI may help to forecast rental income using machine learning models trained on historical leases, vacancy rates and tenant churn. By analysing utility bills, maintenance records and third-party benchmarks, operating expenses can be estimated. However, a lack of detailed public information on commercial leasing and incentives and a lack of details around structural repairs and maintenance are the main barriers to AI in commercial valuations. While the commercial valuation sector highly values data and may be reluctant to share that data with AI, a roadmap to future capability development is essential today. With other technology, such as blockchain, it may well be possible that the source of the data can remain secure with human players, while AI can assist in the analysis and application. As in the past, the valuation industry should embrace technology and work alongside it to remain relevant. Valuers should consider being engaged early in the development process of AI to ensure ethical behaviour and that valuation practices and standards continue to be upheld. Expanding AI's capabilities to value commercial, industrial and special properties, etc., complex assets could be a further study.

The innovative research directions of AI in property valuation hold the potential to revolutionise how assets are valued, offering improved accuracy, efficiency and adaptability in a fast-changing world. Automated data acquisition from disparate sources (e.g., online transaction records, social media sentiment analysis, government reports, satellite images) and the seamless integration of these datasets into AI-driven valuation models for more comprehensive and accurate valuation outputs are challenges, but also opportunities for future research. It is recommended to explore the integration of AI and blockchain to create immutable, transparent and verifiable valuation records that could be particularly useful for data transparency. In addition, investigating AI models that integrate ESG factors into the valuation of assets is an innovation opportunity, allowing for dynamic valuation updates as ESG criteria change and as sustainability regulations evolve. Furthermore, policies to regulate ethics considerations and monitor AI applications in the valuation industry are required for further study.

To conclude, it is understandable that the valuation industry may well feel threatened by the advancement of AI. It should also be noted that the industry has undergone changes for several decades now. At some point, valuers progressed from using formulas to using financial calculators, computer-generated spreadsheets, relying on the internet and using commercially available property databases and software. It is unlikely that valuation as a tool in commerce will always be required; it is a matter of how valuers will again face change and the ability to adapt.

Discussion questions

1 Do you think there is enough readily available data to produce reliable AI-generated valuations?
2 Who would be liable for an AI-generated valuation that may cause a financial loss?

Acknowledgements

The authors appreciate the valuation experts participating in and supporting this research survey and discussions.

Further readings

Benhamou, Y., & Ferland, J. (2020). Artificial intelligence & damages: assessing liability and calculating the damages. SSRN.

Čerka, P., Grigienė, J., & Sirbikytė, G. (2015). Liability for damages caused by artificial intelligence. *Computer Law & Security Review*, *31*(3), 376–389.

Topraklı, A. Y. (2025). AI-driven valuation: A new era for real estate appraisal. *Journal of European Real Estate Research*, *18*(1), 105–120.

References

Abbas, T., & Rinaudo, O. (2025). Bridging fair AI and machine learning in real estate: Ensuring transparency in property valuation and investment.

Abidoye, R., Ma, J., & Lee, C. L. (2021). Barriers, drivers and prospects of the adoption of artificial intelligence property valuation methods in practice. *Pacific Rim Property Research Journal*, *27*(2), 89–106.

Alsahan, I. M., & AlZaidan, Z. I. (2024). Unleashing the power of artificial intelligence in real estate valuation: Opportunities and challenges ahead. *Journal of Artificial Intelligence General Science (JAIGS)* ISSN: 3006-4023, *3*(1), 89–96.

Alsawan, N. M., & Alshurideh, M. T. (2022, November). The application of artificial intelligence in real estate valuation: A systematic review. In *International Conference on Advanced Intelligent Systems and Informatics* (pp. 133–149). Cham: Springer International Publishing.

Amorin, J. (2024). Navigating ethical frontiers: Real estate appraisal in the age of artificial intelligence. *Appraisal Journal*, *92*, 41–58.

Behera, I., Nanda, P., Mitra, S., Rath, S., & Senapati, S. (2023). Analyzing market volatility: Big data's insight into macroeconomic indicators. In *2023 International Conference on Data Science & Informatics (ICDSI)* (pp. 223–237). IEEE.

Braun, V., & Clarke, V. (2006), Using thematic analysis in psychology, *Qualitative Research in Psychology*, *3*(2), 77–101.

Cheung, K. S. (2024). Real estate insights: Establishing transparency–setting AI standards in property valuation. *Journal of Property Investment & Finance*, *42*(4), 406–408.

Chiappetta, A. (2023). Navigating the AI frontier: European parliamentary insights on bias and regulation, preceding the AI Act. *Internet Policy Review*, *12*(4), 1–26.

Guest, G., Namey, E., & Chen, M. (2020). A simple method to assess and report thematic saturation in qualitative research. *PLoS One*, *15*(5), e0232076.

Krivorotov, G., & LaCour-Little, M. (2021). AVMs versus appraisals in refinances: evaluating anchoring aias through relative risk identification. Available at SSRN 3603327.

Krivorotov, V., Kalina, A., Tretyakov, V., Yerypalov, S., & Oykher, A. (2020). Assessment and improvement of competitiveness of industrial complex. Competitiveness Review. *An International Business Journal*, *30*(5), 611–633. https://doi.org/10.1108/CR-01-2019-0011

Lee, C. L., Yam, S., Susilawati, C., & Blake, A. (2024). The future property workforce: challenges and opportunities for property professionals in the changing landscape. *Buildings*, *14*(1), 224.

Liljegren Eriksen, R., & Müller, J. (2024). Using historical sales data to enhance the accuracy of AVMs in real estate finance (Master's thesis, NTNU).

Lucas, S. R. (2014). Beyond the existence proof: ontological conditions, epistemological implications, and in-depth interview research. *Quality & Quantity*, *48*, 387–408.

Mathotaarachchi, K. V., Hasan, R., & Mahmood, S. (2024). Advanced machine learning techniques for predictive modeling of property prices. *Information*, *15*(6), 295.

McCord, M. (2022). A Review of the methods, applications, and challenges or adopting artificial intelligence in the property assessment office. Retrieved on 10 October 2025 from https://pure.ulster.ac.uk/en/publications/a-review-of-the-methods-applications-and-challenges-or-adopting-a

Nguyen, D. (2019). Automated valuation models (AVMs) in the commercial real estate market (Master's thesis).

RICS (2024), New RICS Global Red Book updates mandatory standards for data management technology and ESG, Accessed on https://www.rics.org/news-insights/new-rics-global-red-book-updates-mandatory-standards-data-management-technology-esg-valuation.

Stang, M. (2023). Real estate valuation in the age of artificial intelligence–modern machine learning algorithms and their application in property appraisal (Doctoral dissertation).

Stoyanov, S. (2024). Using artificial intelligence to improve the efficiency of the market valuation method. *Finance, Accounting and Business Analysis*, *6*(2), 217–227.

Szumilo, N., & Wiegelmann, T. (2024). Real estate insights AI: Real Estate's new roommate–the good, the bad and the algorithmic. *Journal of Property Investment & Finance*, *42*(2), 211–217.

Topraklı, A. Y. (2024). AI-driven valuation: A new era for real estate appraisal. *Journal of European Real Estate Research* *18*(1), 105–120.

Umar, J., & Reuben, J. (2025). Fair AI in real estate and insurance: Overcoming bias and improving transparency in machine learning models.

Wei, C., Fu, M., Wang, L., Yang, H., Tang, F., & Xiong, Y. (2022). The research development of hedonic price model-based real estate appraisal in the era of big data. *Land, 11*(3), 334.

Yiu, E. C. Y., & Cheung, W. (2024). How to bring AI-enabled valuations to meet professional duty of care standards? Available at SSRN 4720254. https://doi.org/10.2139/ssrn.4720254

Zurada, J. M., Levitan, A. S., & Guan, J. (2006). Non-conventional approaches to property value assessment. *Journal of Applied Business Research, 22*(3), 1–14.

Section 2

Applied digital innovations in valuation practice

Chapter 6

Public sector valuations

Current and emerging technologies, and industry readiness

Elise Wallace, Adam Wallace, David Logue and Matthew Moorhead

Introduction

This chapter examines technology use across the valuation of Australian public sector assets: land, buildings and engineered infrastructure and considers how emerging tools affect quality, cost and audit outcomes. Because these assets predominantly appear on public sector entities' balance sheets, they must be revalued annually for inclusion in publicly released financial statements in compliance with the requirements of the Australian Accounting Standards Board (AASB, 2023). The valuations inform key inputs within financial statements, which become publicly available documents. While it is necessary for valuers to adhere to the International Valuation Standards Council's (IVSC) relevant International Valuation Standards (IVS) when undertaking these valuation assessments, it is the AASB that prescribes how the assets are measured within the 'Fair Value' context and the methodology disclosure reports are provided (AASB, 2023, para. 9). In practice valuers can play an interpretive role for clients and key stakeholders in explaining and contrasting the differences and similarities between the AASB and the IVS, including definitions and requirements which show considerable variations in different contexts.

Infrastructure assets, land and particularly the networks of roads (and bridges), water, sewer and stormwater provide fertile ground for analysis since, despite rarely transacting in open markets, they generate large volumes of data. The principles examined herein extend to the broader public sector infrastructure assets portfolio. In this way, the chapter speaks directly to the book's wider theme of digital transformation in valuation practice. The chapter is organised in three parts: first, inspection hardware; second, data-centric and artificial intelligence (AI)-driven tools; and third, governance, procurement and ethical considerations.

Every valuation performed for a government entity under the AASB is subsequently audited, typically by a state audit office or an external audit firm engaged on its behalf. The work may take weeks or, for large portfolios, even years, depending on the size of the asset register. Government agencies

DOI: 10.1201/9781003539254-8

collectively control thousands of land parcels, buildings, roads and other infrastructure assets, whereas Local Government Authorities' (LGA) property and asset portfolios can vary, ranging from the extensive portfolio of large LGAs such as Brisbane City Council to small rural shires with modest holdings. Many entities, therefore, adopt a rolling programme in which only part of the portfolio is valued comprehensively (that is, with physical inspection) each year. Efficiency is paramount, as valuers must meet statutory deadlines and still deliver high-quality and cost-effective outcomes for ratepayers and taxpayers.

Public sector portfolios span many asset categories, each of which demands specialised expertise. Asset portfolios are therefore disaggregated into *asset classes* such as:

- Land
- Buildings (including investment properties)
- Site Improvements (sometimes referred to as Other Structures)
- Water and Sewer Assets (such as treatment plants, pumps, pipelines and pits)
- Stormwater (surface and subsurface assets, drainage pits)
- Open Space Assets, park infrastructure/embellishments
- Community facilities (libraries, swimming pools, community halls)
- Roads and Footpaths
- Bridges and tunnels
- Aerodromes
- Artwork
- Book/Library Collections
- Plant and Equipment
- Landfill assets (for example, leachate ponds, cells, transfer stations)

Even within a single class, the asset mix is highly diverse and provides complications that valuations of more regular assets do not necessarily encounter. For example, in many cases, the inherent value lies in components that are underground or otherwise partially or completely hidden from view. Experienced valuers within this space would likely have engaged with a vast array of more peculiar and specialised assets, including satellite dishes, observatory towers, aeroplanes, boats, stone churches, stadiums, marine walls and defence equipment. Moreover, there is significant variation in the geography, terrain and climatic factors these assets exist in and are exposed to. Because of this heterogeneity, technology plays a different and often more nuanced role than it does in other valuation contexts, particularly in relation to data capture and specialised software.

Government agencies are sometimes stereotyped as 'late adopters' of innovation and technology. Notwithstanding this, many benefit from private-sector innovation through engaging external consultants using contemporary technology, providing efficiencies and more nuanced analysis developed through ongoing exposure to higher volumes and more diverse types of assets. Because technology evolves comparatively quickly, reviews of technology carry a more elevated risk of rapid ageing. Accordingly, this chapter focuses on broad trends rather than endorsing specific vendors or platforms.

Numerous stakeholders rely on public sector asset valuations, including internal finance teams, asset managers, engineers, accountants, auditors, executive managers, asset custodians, consultants engaged by stakeholders and, ultimately, the public. In recent years, valuation and asset management functions have become increasingly intertwined, promoted in part by state audit offices and government agencies. The Australian Local Government Association's (ALGA) National State of the Assets (NSoA) Report evidences this trend as it seeks to measure sustainability, future-proof communities and monitor the performance of Local Government Authorities (Australian Local Government Association ALGA, 2024, p. 4). It states,

> This report is based on data collected from approximately 85% of Australia's Councils across eight different categories of infrastructure assets.
>
> ...
>
> The task of monitoring the performance of local government assets in all categories is resource intensive, requiring skilled professionals to make careful judgements on the remaining life of a vast array of materials.
>
> The report highlights the importance of reliable and credible data to support decision-making and the critical role of asset management planning and long-term financial plans in every council.
>
> (Cook, 2024)

In addition, rather than various government agencies undertaking asset class projects requiring assessments that produce information that is highly specific or comparatively fragmented (such as an individual condition assessment project as opposed to a valuation), it is more cost-efficient to have a single inspection serve multiple purposes (for example, valuation, accounting compliance, condition inspections, asset management metadata updates, defects and maintenance reporting, budgeting, forward planning, spatial data updates and photo documentation). To date, this process has progressed organically because of the way valuations report remaining service potential and asset condition, which is a critical input to the Current Replacement Cost (CRC) method. This method is advantageous when the valuer is applying a condition assessment, which is a component of the process of assessing the remaining useful life of an asset, and further allows application to other areas of the entity, such as budgeting, maintenance and sustainability ratios.

The CRC method is applied in the majority of assessments for public sector entities since the assets are thinly traded and where they are, are rarely, if ever, at a commensurate highest and best use or in an arm's length transaction.

Thus, there is typically a dearth of reliable market evidence when undertaking valuation assessments of assets such as roads, toilet blocks, arts centres, sewerage treatment plants, cemeteries and the like. However, these circumstances do not preclude using traditional market-based methods (which are, in fact, preferred). Ultimately, the relevant methodologies are selected by the valuer based on the availability of observable inputs such as market data and the characteristics of the asset, as set out in the *Queensland Treasury Non-Current Asset Policy for the Qld Public Sector (NCAP) – No. 3 'Valuation of Non-Current Assets'*.

Fair value under the CRC method relies on several data-intensive inputs, for example:

a) Input rate (for example, replacement cost per m^2) unit
b) Quantity (for example, area, length, capacity, etc., or another unit)
c) Economic Life (total useful life of an asset, for example, years)
d) Remaining Service Potential score (for example, physical condition assessment incorporating obsolescence allowances)
e) Remaining Service Potential (for example, remaining useful life, (years))

Valuation is recognised as being 'not an exact science'; reasonable variation is to be expected in some aspects, particularly market interpretation and obsolescence assessment. For the CRC method, the assessment of remaining service potential, due to the difficulties of measurement, can produce comparatively fewer specific outcomes. Conversely, the assessment of economic life, unit and input rate is comparatively more objective due to the ready availability of observable data. The inputs are important due to the role technology plays in the application of processes. Furthermore, technology is infiltrating the entire valuation process, from inception to audit, with the latter relying heavily on data and analytics. Public sector valuers and auditors increasingly rely on a growing suite of hard and soft technologies, each of which is being continually refined and has improved the speed and rigour of their work. Public sector agencies often adopt hardware-centric inspection technologies several years after private industry, predominantly due to the more complex procurement processing, budgeting and risk-management cycles they are compelled to follow. Data collected during major 2023–24 drone and 360-camera rollouts in Queensland and Victoria illustrate both the scale of that lag and the benefits once the tools are operational, as this chapter will now explore.

Hard (physical) technology

Several initiatives are presently in use that streamline data collection for remote and complex assets. The tools used in the field inform both the operator and the data system engaged in the background. Bluetooth laser measuring devices, high-quality cameras and tablets are now essential and commonplace amongst valuers in this space and in traditional valuation offices. The following sections outline the key technologies and describe how they improve safety, accuracy and efficiency in public sector valuation.

Drone-based inspection technology

Drones are a relatively recent tool that provide valuers safe access and improved observation abilities, particularly for identifying and assessing the condition of roofs and roof-mounted infrastructure and elevated structures. High-resolution imagery, video and LiDAR data can be captured quickly and geo-referenced for later analysis. Current widely available models perform key tasks, including internal wall imaging (exposing thermal or pest activity), detecting radiation and providing 3D modelling and measurement, providing far greater efficiencies and information analysis for valuers in both the inspection and report-writing stages.

The valuation industry has reported mixed reviews on this technology. Key positive attributes include the ability to reach places difficult or harmful for a human, including under bridges, the roofs of buildings, communication towers or other parts of infrastructure assets. Well-resourced local governments are increasingly adopting this technology. Supported by specialised software, drones can generate georeferenced 3D models of assets that add significant value to asset management workflows. However, higher capital costs, restrictive licensing requirements and regulatory compliance remain substantial barriers for many organisations that wish to bring the capability in-house. Uptake of technology and services using this technology has been varied, particularly where the quality of the technology's output was less than expected (for example, image detail). Notwithstanding this, consultants who have a greater command of these technologies are growing in popularity and demand in both the private and public sectors, further enabled through recent advancements in the quality of the outputs.

360-degree cameras

Handheld 360-degree cameras capture immersive, three-dimensional imagery and can replace traditional still photography on site. Geo-tagged footage supports desktop review and audit queries long after the inspection date. An inspector can walk through a building gathering necessary room data, which is GPS logged, and create a full spatially referenced layout of an asset. Combined with drone technology capabilities to reach inaccessible areas, a comprehensive 3D model can be pieced together of the entire building, leading to the creation of 'Digital Twins'.

These functions have proven useful from both an asset management and valuation perspective. When combined with suitable machine learning models, the imagery can auto-generate building-information-model (BIM) files, significantly reducing manual drafting time. For example, practitioners using 360-degree cameras on inspections for stormwater and drainage assets (pits and pipes) overcame restrictive conditions where pit width or damaged stormwater pit lids prevented access to assets, in conjunction with inadequate light. Using 360-degree cameras enabled capture of the various components of the pits and pipe entries, enabling a better assessment of condition, identification of defects and ultimately assessment of the value of these assets. More importantly, it also reduced work health and safety risks for practitioners while also avoiding the additional confined space entry risks.

Virtual reality and augmented reality

Virtual reality (VR) and augmented reality (AR) applications allow valuers and asset managers to 'walk through' complex structures virtually, overlaying live defect data on the digital twin. Head-mounted displays can be taken into the field or used in the office for audit review. AR will continue to lend itself to more practical situations for an inspecting valuer. It is an emerging space within the industry, with many larger-scale technology firms working on AR solutions, although many continue to be in the development stage. At the time of writing, there are no documented real-world applications focused specifically on asset inspections. Current concepts use 'glasses'-style devices, which are lighter than typical VR units that provide information on the screen as the valuer proceeds through an inspection. This may provide an opportunity for significant improvement to the inspection process, combined with machine learning (ML) models to identify and verify common asset elements, along with defect and condition identification (see also thermal/moisture imaging below). When AR equipment becomes more widely available and cost-effective, pairing it with machine learning models will further enhance physical asset inspections. Integrating thermal-imaging and moisture-sensing cameras into AR devices would represent an especially significant advance. Current VR systems rely on bulky head-mounted displays that are impractical for on-site inspection and valuation. When linked to accurate 3D building models, however, VR can add value as a desk-based visualisation tool for asset management planning. Industry uptake remains low, and practical use cases in valuation and asset management are still limited.

Tools for additional condition assessment

Although these tools are still used infrequently in valuation, they are increasingly valuable to asset management workflows—especially when detailed condition assessment and lifecycle analysis are required for critical assets. Specific examples are described as follows:

- **Acoustic-emission testing** – rarely employed in valuation, but highly relevant to asset management. Strategically positioned sensors record stress waves within a structure, delivering continuous data on external loads and structural integrity.

- **Thermal-imaging cameras and moisture meters** – handheld devices already common in building inspection. Incorporating them into valuation site work deepens condition assessments by giving stakeholders subsurface data. When combined with AR/VR visualisation and drone platforms, these sensors can 'see' inside wall cavities, pinpointing leaks and other defects that are invisible to the naked eye.
- **Building sensors and building management systems (BMS)** – centralised computer systems that are used to manage, monitor and configure elements and services within a building, such as HVAC (Heating, Ventilation, and Air Conditioning), lighting, power, fire and security systems. Use within the asset valuation sector has not yet been observed. However, these systems can have an important role in the asset management process and to date have been typically restricted to more high-profile buildings with larger budgets.

Soft (data) technology

The data-technology landscape is evolving rapidly across numerous facets of the valuation profession, making it challenging to stay abreast of every development. Nevertheless, Australian Government agencies—whether in-house or via external contractors—are actively trialling the following data- and software-based solutions.

Geographic information systems (GIS) software and other spatial tools

One of the first questions a valuer must answer is 'where is the asset and how do I access it for an inspection?' This is of particular importance for government infrastructure in remote locations, sometimes in uninhabited parts of the country and in potentially dangerous climatic and geographical conditions. Using GIS and spatial tools is of significant aid in this important first step. Valuers with GIS and mapping expertise are an increasingly important skillset for large-scale portfolio valuations. Most government agencies employ their own specialist GIS teams, and valuers need to have sufficient skills to access, review and work with these organisations' spatial data and the systems that they use.

GIS systems enable the visualisation, analysis and editing of spatial data. Platforms range from sophisticated aerial-mapping suites such as ArcGIS to widely used and more simplified web- and mobile-based applications such as Google Maps and Google Earth. GIS organises information into map 'layers', each georeferenced to precise GPS coordinates and accompanied by rich metadata. Across Australia, all tiers of government now release property-related GIS layers through open portals that are freely accessible to the public, typically used on local government planning and development websites and some state government websites, such as the Queensland Government's Queensland Globe. Property-relevant layers typically include:

- Base lot and plan data, containing lot, section and survey plan numbers, tenure, land size and local government areas, while also visualising the property boundaries.
- Zoning and town planning layer data.
- Environmental Overlays.
- Flooding Layers.
- Contaminated lots.
- Easements.
- Aerial/satellite imagery.
- Topography and contours.

Having this data digitised and openly available for the public to use is very important to the property valuation process, particularly for land valuations, which aids in simplifying the task for valuers. Being able to visualise the land and its many different attributes provides critical insight into the characteristics (and ultimately value) of the land. The visual process also allows for sales analysis in proximity to the land parcel, as well as comparative analysis of multiple lots within a similar location, along with the other land characteristics such as the layers noted above, all providing greater consistency and quality assurance to the valuations. Moreover, as valuers are now increasingly required by legislation and professional practice standards to report and consider the effects of ESG (Environmental, Social, and Governance) in valuations, the information in these spatial tools can contribute to the detail required for this reporting. Private service providers now supply regularly updated satellite imagery to many of the major urban areas of Australia. These tools are typically clearer and more regularly updated than the free versions, for example, Google Maps, and allow the valuer to also view historic imagery, providing a useful snapshot of changes to an asset over time. Image resolution continues to improve, providing higher definition, enhanced measurement capability and more innovative tools to work with the images and data, which may include single assets.

The increasing integration of valuation and asset management means that precise asset location and mapping are as critical as the value itself. Inspection software commonly in use at present is accessible via mobile device inspection tools, for example, tablets with Android and Apple functionality, which use a mix of satellite imagery and some proprietary GIS layers specific to the needs of valuers. To date, they have proven to be essential tools for valuers conducting field-based assessments to accurately identify and record the necessary data and rely upon it to make relevant value-based assessments of the assets.

Machine learning and artificial intelligence (AI) with AI-powered inspection software and predictive maintenance modelling

The AI definition used by the Australian Government in their September 2024 policy for responsible use of AI in government is derived from the Organisation for Economic Co-operation and Development (OECD):

> An AI system is a machine-based system that, for explicit or implicit objectives, infers, from the input it receives, how to generate outputs such as predictions, content, recommendations, or decisions that can influence physical or virtual environments. Different AI systems vary in their levels of autonomy and adaptiveness after deployment.

ML models are important verification tools to identify common asset components, defects and conditions from the field inspection photos and pre-populate data entry. They can complement and accelerate the inspection process while reducing human error. Further privacy-focused technologies, such as facial blurring and data anonymising, are expected to ease public concerns about breaches of privacy.

Adoption of this technology appears to be at a comparatively low level, with firms in an exploratory stage. A number of field-based challenges still exist, which have demonstrated that adoption of AI is still not likely in the short term, despite the anticipated benefits.

The ongoing maintenance requirements and associated costs currently outweigh the benefits of reduced subjectivity that existing methods include. For smaller firms, the business case shows that the adoption of AI is not yet feasible. As the technology matures, it is anticipated that it will learn and apply additional valuation inputs such as automated asset categorisation; for example, recognising a public amenities block and assigning the relevant unit rates applicable to the gross replacement cost. Ultimately, faster and more consistent processing should deliver more reliable valuations.

Predictive AI tools for market valuation assessments are already available and have gained traction in traditional valuation segments such as mortgage lending and statutory valuations, although they are still not widely adopted due to comparatively less reliable accuracy. This improves where transaction volumes are high and data quality is robust, conditions most commonly found in the residential sector, which is both the most frequently traded asset class and the domain in which mortgage valuation platforms capture the greatest depth of data. Notwithstanding this, the accuracy of predictive AI tools in these value models is still comparatively poor.

In the United States, several commercial platforms now apply predictive analytics to statutory valuation analysis with the transparency and benefit of seeing (a) what evidence has been utilised by the AI, as well as (b) the adjustment factors that have been applied to inform the assessment. This way, the valuer can not only see the decision-making process but can weigh in on the outcome. In one article, the term *'AI infused solutions'* was mentioned, and we see more and more that this is likely an apt description of striking the balance between human and AI relationships for professional services:

> One of the key benefits of the AI-infused data modelling platform is that it has revealed which data has most influence on the accuracy of predictions, allowing . . . to focus its efforts on collecting the right data to feed into the . . . platform. Variables such as building area and land area have great influence . . . has also been able to leverage other locality factors that people consider ahead of buying property, such as driving distance to railway stations in urban areas.
>
> We use the machine learning to inform us about errors with our models . . . It is significantly increasing the confidence in the outputs we're able to put forward for councils. We are making sure people aren't being over charged and certainly the efficiency level it's helping us achieve is dramatic.
>
> While . . . has not replaced its own valuation model with the AI platform, it has compared the values predictions of each. In a trial the company found that the AI model was more accurate in 75 per cent of cases. .
>
> (Microsoft News Center, 2019, paras 5–6)

In Australia, there are similar initiatives, including one being trialled in local governments across New South Wales, based on an investment by the New South Wales Government to support their local planning systems. Further, a collaboration project between the PEXA Group and the Australia Property Institute (API) is based on a programme to revolutionise valuations and analytics. This was initially conceptualised by the University of New South Wales and FrontierSI and is based on six years of research and development through a pilot programme across metropolitan Sydney and Brisbane, and the software features a customisation module specially for Valuers General in Australia (Value Australia, n.d.).

Automated report generation

Automated reports based on the data recorded digitally by the valuer during the inspection and valuation process are presently in use and allow for the rapid creation of professional documents to expedite report completion and delivery to clients. This is in contrast with long-standing and established practices requiring manually typed reports by administration staff, deciphering handwritten notes from a valuer and requiring good relationships with photo printing companies to return physically printed photos which were adhered to a report. Automation of these activities has been increasingly available for some years, and many technological products continue to evolve. The desired outcome of both client and firm is to generate reports in 'real time' once an inspection has been completed. AI is expected to increase the accuracy of these reports, reducing human input errors and prepopulating data automatically, as well as providing quality assurance reviews.

Mobile apps (with devices)

There are a number of service providers that provide mobile applications which allow tailored data entry forms, photography and documentation during the inspection process for valuers in the field. The essential technologies are offering increasingly advanced GIS/spatial functions to assist in identifying and measuring assets, and ML integration is increasingly implemented to assist the inspection process.

Programme management to improve workflows

There are many project/programme management tools on the market, with cross-device compatibility allowing in-office and in-field communication, with extensive capabilities such as team- and project-based communication tools, real-time notifications, project-specific document repositories, resource/staff allocation, budget planning and time tracking. These tools and processes are becoming essential to the workflow in large portfolio-style valuation projects. With many moving parts within a valuation practice and concurrent (large portfolio) projects that can span many months, it is crucial for management to have access to these tools.

WHY EMBRACING TECHNOLOGY IS IMPORTANT FOR PUBLIC SECTOR VALUATION

A speaker at a 2025 property industry conference in Australia stated that *75% of Australian workers are using AI at work, and 78% are bringing their own AI to work* (D. Hookham, personal communication, November 21, 2024). Apart from statutory valuations and individual workers who might dabble with AI tools to assist them with repetitive tasks or report writing, there appears to be a limited amount of ***internal*** AI systems being implemented for valuations of public sector assets, and a greater reliance on external contractors and data experts has been observed. It is clear that the valuation industry is seeking ways to embrace the idea of generative and large language model AI and other such software, as well as advancing hardware to facilitate 'smarter' inspection processes. It is apparent in the valuation industry (and not uncommon in most professional services) that AI adoption is motivated primarily to drive efficiency, compliance and accuracy. Repetitive or mundane tasks within a valuation or administration job are the most obvious candidates for the adoption of AI, notwithstanding the necessity to improve the quality of inputs needed for the valuation of public sector assets. Continuous improvement is paramount, as valuation advice directly influences the financial sustainability of government agencies—particularly since they are accountable to ratepayers and taxpayers.

> The Microsoft Copilot Trial involving over 7,600 staff across more than 60 government agencies from January to June 2024 was carried out with the aim to determine whether integrating AI tools could enhance productivity, improve workflow efficiency, and support staff in their daily tasks while ensuring ethical considerations and governance were maintained. On average, participants saved 1 hour per day on tasks like note-taking and administrative duties, with 40% of that time redirected towards skill-building activities. 69% reported a noticeable improvement in the speed of completing tasks, and 61% felt that the tool enhanced the quality of their work output.
>
> (D. Hookham, personal communication, November 21, 2024)

The tools more broadly adopted by this industry facilitate work by private consultants to government clients, are important to meet the needs of those clients who work on tight financial year cycles, provide better consistency and accuracy in analysis (particularly with projects requiring teams of valuers) and stay competitive. The GIS tools and electronic tablet inspection hardware, along with proprietary inspection software, have significantly increased productivity and have provided a level of accuracy not possible with traditional methods – such as geolocating the assets and providing GIS layers with our inspection data matched so clients can replicate their assets' information on a digitised map. Furthermore, in-house scripts provide the ability to clean (normalise) market or construction cost movement data by removing abnormal and irrelevant records (such as related party, multi-sale, and part-sale) and run this analysis within seconds, rather than (the average) 3–4 hours. Such software can create data collection forms based on each client's needs and can be updated, changed or adopted in real time. All photo, sketch and data information on an asset or its evidence is captured for the asset, and cloud-based storage allows ready access. Further, the platform allows multiple valuers to gather and upload data at the same time and within the same location, without cross-over. From an audit perspective, being able to provide consistency and accuracy for key reporting metrics is essential. Having access to a database of analysed government construction projects generated from years of collecting construction data provides the ability to mix construction cost guide information with real project data and produce support for the rates adopted.

The potential of AI to insert bias into the work is still being understood. Whereas it is well known that poor input information will result in poor information outcomes, considerable efforts are necessary to ensure that only quality, targeted information is provided to the AI in a sufficient volume.

The majority of this quality information is presently held by private consultants, and being trained on public information may not produce the same level of precision when much of the "good stuff" sits within private enterprise. It is foreseeable that the audit process will also become more reliant on AI, using the valuers' data to quickly look for anomalies and outliers or provide comparable data sets to benchmark with others. Large auditing firms like 'the Big 4' and the state audit offices are already investing in this space.

Influence of AI and other technologies on procurement and policy development

In Australia, the public sector asset valuations industry is a comparatively low-volume market segment against traditional valuation work, such as valuations undertaken for mortgage lending, where a far greater number of valuations are undertaken each year. To date, within this market segment, there is limited evidence that AI and other technologies are being used within government agencies' procurement processes, particularly at a large scale. Notwithstanding this, there is anecdotal evidence that AI is being used to summarise, compare and contrast tender submissions to create efficiencies in the evaluation process for key decision makers. Furthermore, evidence suggests a trend of greater standardisation of tender forms and other documentation designed to be suitable for processing by AI. Hence, the emerging pattern demonstrates continued integration opportunities for AI into procurement selection processes that provide rapid production of compare/contrast summaries, risk profiles and compliance checks for tender submissions to enable rapid tender decision-making. Whether, and to what degree, this removes subjective elements from the process is not yet apparent.

The emergence of AI within this space has necessitated a government agency policy response, and it is also a developing area. Government agencies place a high level of trust in and acceptance of the professional capabilities and audited processes of the valuation firms undertaking public sector valuations. This trust is also demonstrated by the financial system and the courts, which rely upon the professionalism of valuation firms. Anecdotal evidence suggests the public sector asset valuation industry continues to carry a degree of scepticism for the adoption of AI in creating reports used in the valuation process and in carrying out the valuation analysis processes undertaken by valuers. This is not limited to only valuations. For example, at the request of government agencies, quantity survey reports carry specific qualifications stating, 'no AI has been used to create this report'. Consequently, policy development within public sector asset valuations generally limits the use of AI in valuations to only administrative or standardised and repetitive tasks.

The following section explores the barriers valuers will possibly face in the future while undertaking valuations of public sector assets using these technologies.

BARRIERS TO ENTRY

At an Australian conference of government finance professionals in November 2024, the room of delegates was asked a series of survey questions around their knowledge and/or implementation of AI technology. Within the survey, 79% of the respondents agreed that AI had a role in optimising their finance operations, though 57% were not using AI technology within their department. When asked what was restricting them from using AI, the key themes in the responses were:

- Not knowing where to begin.
- A lack of understanding.
- Data safety not permitted by the entity's management or IT personnel.
- A lack of policy around its use.

Uptake has understandably been slow in the government sector. However, on 1 September 2024, the Australian Public Service (APS) issued its Policy for the Responsible Use of AI in Government, a move that should accelerate adoption. From an industry perspective, the following challenges are emerging

Procurement and government policy

Typical technology procurements now pass through a five-gate process—strategic assessment, business case, market engagement, full business case and benefits realisation—each gate requiring progressively richer evidence of risk allocation. Clauses most often scrutinised by Crown Solicitors include intellectual property ownership of captured data, uptime service-level agreements for cloud platforms and indemnities for algorithmic misclassification. Together, these frameworks slow—but greatly clarify—the adoption pathway and give Auditors-General confidence that valuation inputs remain defensible.

An abundance of choice and data integration

Currently, practitioners are faced with a proliferation of competing platforms and technical models, and providers are facing difficulty making decisions on the best fit. With the possibility of many different providers, platforms, and datasets being generated, 'all in one' integration of information remains a challenge. The role of data analysts and data technicians will play an important part in effectively managing, maintaining and brokering the right fit. Exploring the marketplace for providers of optimal solutions for government and valuation-specific tasks is a time-consuming exercise and includes sales pitch meetings, reviewing quotes and reading product statements. And it is apparent why many firms are creating their own in-house programmes, despite their challenges and, of course, costs.

Cost

With many of the technologies discussed currently in the 'groundbreaking' phase, early adopters typically pay higher prices in comparison to when wider adoption occurs, especially with new hardware and specialised equipment. The earlier adoption also means that many of the 'bugs' and refinements have not yet occurred, making integration difficult. Bespoke in-house models also absorb a significant amount of time and resources, thus increasing cost, as specific tasks within ML (for example) will require high-quality data sets and therefore require curation and verification. This model is not always viable for smaller consultancies, raising concerns about their long-term sustainability in Australia. Nonetheless, more cost-effective off-the-shelf programmes are likely to emerge—although they may further compound the current 'abundance of choice'.

Data privacy and security

Apart from data stored internally within government agencies, with the abundance of technologies and services emerging, data has the potential to be stored in many different third-party services and technology providers. Managing where this data is held, who has access, and how well the machines are secured (both physically and through software and regular updates/patching) poses risks to both government agencies, contractors and third parties who store the data. It is expected that tenders will need to evolve to ensure more safeguards are applied to data security. The Australian Government has adopted the Essential 8 model for cybersecurity, and state and local governments are expected to raise their maturity levels accordingly (Australian Cyber Security Centre ACCS, n.d.). Private-sector organisations that handle government data will therefore need to demonstrate compliance with the same controls.

With 'cloud' data storage increasingly becoming a predominant feature/focus for emerging AI technology, questions arise as to where the data is located. Government agencies now routinely ask contractors to warehouse all data within Australia, and many procurement panels list onshore storage as a preferred or mandatory criterion. Privacy remains a critical consideration, and tools that anonymise data are becoming increasingly important. Automatic facial blurring in imagery, for example, is an important first step towards a privacy-centred approach that new technologies should embrace to gain broader public acceptance.

This issue also encompasses operational risk: if a government agency or its contractor were to lose core data through inadequate safeguards, the time and cost of recovery could be severe. Rigorous backup and disaster recovery policies must therefore be treated as a top priority by IT personnel. Yet instances of unrecoverable data loss still occur within government departments for a range of reasons. Additionally, social engineering and/or employee malice are considerations in the protection of data, with several cases experienced in recent years within the valuation industry. All too often, the media is peppered with stories of mass data breaches as a result of breaches of security—whether physical security, software security built into systems and databases, or even the human element by which data is leaked via bad actors manipulating employees or where disgruntled employees have maliciously damaged company technology, property, data and clients.

From a property inspection perspective, it is anticipated that AI-assisted tools will receive mixed public acceptance. Many public sector sites already prohibit photography for security reasons, making the use of 360-degree cameras and drones even more sensitive, including culturally sensitive areas and communities. At this stage, it would be difficult to imagine how this issue can be carefully implemented and communicated to the community or individuals affected, especially in homes and facilities where there may be an anti-government and/or anti-technology bias. This bias may be exacerbated if the present underlying public and government concerns of privacy breaches and the risk of privacy or personal information manipulation by large corporates, 'big-tech' or state-based entities continue. This continues to be an evolving space with the technology upon which it is based.

Regulatory compliance and risk management

A major concern for government agencies is regulatory compliance and risk management, which has slowed the adoption of this technology. Increasing focus will be placed on policies and procedures surrounding data and management, as well as ethical and safe use. Policy manuals such as Business Continuity Plans, Data Disaster Recovery Plans, Data Protection Privacy Policies, and Data Security Policies will all need to be implemented (if not already) and regularly updated. The Policy for the Responsible Use of AI in Government, released by the Australian Federal Government, took effect on the 1st of September 2024. Individual state and local entities, however, will need to develop their own frameworks and implement/train staff on the expectations around this. The evolution of change management in government is a wide topic in itself, worth exploring. Regulatory compliance and risk management must not be scaled back as the risks continue to evolve and increase. For example, an AI large language model (LLM) summarisation tool can generate 'hallucinated' data points. All AI-generated outputs, therefore, undergo a manual credibility check before they are exported to the valuation report template.

Skills gaps

Looking ahead, universities, TAFEs and professional bodies will need to embed LiDAR processing, drone-inspection workflows and AI ethics into curricula and Continuing Professional Development programmes to ensure a sustainable talent pipeline for public sector valuation work. Software integration into organisations requires a commitment of resources and time, and with AI, it may be even more imperative to have well-trained and skilled users to be stewards of the data which the government holds. Human error is often a major cause for data loss; under-skilled staff have the potential to cause damage, loss and corruption to their important data sets. Initial as well as ongoing training will need to be a cultural norm. Blind reliance on the assumed accuracy of AI platforms is equally of concern, and students must also be cognisant of the need for the development of critical thinking skills within the AI space.

Unrealistic expectations

Exaggerated marketing claims have fostered unrealistic expectations about AI's capabilities, leading some users to rely too heavily on its outputs without adequate professional scrutiny. This issue flows directly into the broader ethical considerations surrounding AI adoption. At the November 2024 Local Government Finance Professionals Conference in Queensland, consultancy Component 6 highlighted several key ethical themes, which are expanded upon below (Boyd, 2024).

Algorithmic bias

AI systems can unintentionally perpetuate biases found in the training data. At a recent conference held by Central Queensland University, this was explained in the discussion of a trial where AI was 'taught' all areas of dark matter relating to the human condition. Upon being shown an ink blot artwork, the AI 'saw' a man being electrocuted and facing certain death (Boyd, 2024). That certainly wasn't what the audience saw, and demonstrated how biases accumulated through training can influence the outcome. A valuer's experience and understanding of the market participant, which can be influenced by complex 'human' reactions, is critical. Data is important, but it can also sometimes be only part of the picture. With the recognition of algorithmic bias, mitigation techniques are being sought to improve the quality, such as bias detection tools and fairness-aware machine learning. The continual training and improvement will lead to improvements in the next generation of models.

Transparency and explainability

Not understanding how AI decisions are made can lead to 'black boxes'. As valuations need to be explainable, well-referenced and transparent, not just for audit purposes but for all stakeholders, experiencing this outcome may undermine any trust put into our assessments. Valuers must seek technologies that allow their experience to be incorporated into any assessment that is undertaken (so AI infused tech) and develop policies around the use of this tech so that valuations are transparent and methodologies (valuation and AI) are explained.

Accountability

Who is watching the watcher? If AI is making the decisions, who is monitoring the quality and accuracy of those decisions and where does the responsibility for those decisions lie? Is there an over- or under-reliance on certain technologies within the valuation process? Valuations are being provided so government agencies can rely on this data for maintenance and operational budgets, asset management planning, financial and strategic purposes, as well as reporting to their voters/ratepayers/taxpayers. It is essential that the valuations are not left entirely to the hard and soft technology solutions (hence the reason the term 'AI infused software' fits a lot more neatly in this space), and the equilibrium between the machine and 'valuers opinions based on experience' needs a respected balance. Furthermore, it is important for auditors too, because if 'continuous, data-driven auditing' becomes encumbered and failures occur as a result, this will have enormous repercussions for the public. Auditors apply best practices to understand how valuations are performed under the accounting standards. It would be a disappointing outcome if there were an over (or misguided) focus on potentially contaminated data.

Trust and scepticism

This is a contested, volatile and evolving space, and therefore, dangerous to rely solely on technology without human review and intervention. There is currently considerable reporting on the inaccuracy of current AI models and erroneous (or at least unreliable) outcomes, decisions, and responses (Knight, 2021; University of Maryland Libraries, n.d.; Marwala, 2024; OpenAI, 2025). Moreover, continued community backlash may eventuate in escalated use of AI by governments in areas of personal property. It is therefore unlikely that technology will completely replace the role of the human inspector.

Conclusion

Technology has seen increasing adoption in the valuation space in recent years, albeit from a low base. What is now clear is that it will remain increasingly critical in the valuation space. For public sector asset valuations, technological advances can most effectively be applied within the inspection support and the data-heavy aspects of the process, particularly the capture and analysis of inputs to assessments using the Current Replacement Cost method. As an established, predominant methodology applied to the valuation of specialised assets, most prominently infrastructure assets, its veracity relies on accurately identified, classified and quantified data and robust analysis. Greater efficiencies are likely for market value methodologies, including fast-paced analysis of market trends using bulk transaction data, consistency and better processes for selecting and analysing sales comparisons, as well as the grid adjustment matrixes that underpin them.

Advances in this area are anticipated to more readily lie in improved automation using more sophisticated analysis tools such as capitalisation and discounted cash flow (DCF) templates, and in reporting, including variations in communication. One of the most significant foreseeable benefits in the not-too-distant future is superior quality assurance and data testing. Using databases and 'fed' information, valuation systems are anticipated to have greater benchmarking capabilities, error identification and the ability to rapidly query such anomalies. Auditing time will thus be reduced, resulting in a faster and more accurate reporting process, resulting in superior product and service delivery borne through effective integration of valuation experience and technology application. Within this, it will be necessary for government agencies and their consultants to develop robust policy and governance frameworks in close consultation with communities, particularly those that have limited tolerance for privacy erosion, to ensure the safe and accepted integration of these technologies. These frameworks should also address stringent data security measures and targeted staff training.

In summary, the valuation of public sector assets in Australia now sits at the intersection of advanced inspection hardware, data-driven analytics and increasingly stringent governance expectations. Drones, 360-degree imaging, more advanced GIS layers and AI-assisted models already enhance speed, safety and consistency, while emerging tools such as AR-enabled inspections, automated report generation and predictive maintenance algorithms promise further gains. Yet technology alone is not a panacea: robust data standards, transparent "human-in-the-loop" workflows, onshore cyber-secure storage and disciplined staff training remain essential to guard against bias, privacy breaches and operational risk. By pairing these safeguards with a measured, evidence-based embrace of innovation, valuers, auditors and asset managers can deliver more reliable, auditable and cost-effective outcomes which ultimately strengthen the financial sustainability and culture of trust of the public entities that serve Australian communities.

Discussion questions

1 What do you think these mean for the future of the valuation industry: opportunities, threats, weaknesses and strengths?
2 What do you need to look at to be best equipped for being a valuer or in the valuation industry, for example, what software or hard skills will best equip you?
3 Where do you believe the government should regulate or draw the line with the integration of AI, both internally and with its service providers?

Resources for further reading

https://www.digital.gov.au/policy/ai/policy
https://www.finance.gov.au/government/public-data/data-and-digital-ministers-meeting/national-framework-assurance-artificial-intelligence-government
https://www.dta.gov.au/

References

Australian Accounting Standards Board. (2023). Fair value measurement (AASB Standard No. 13). https://standards.aasb.gov.au/aasb-13-dec-2022

Australian Cyber Security Centre. (n.d.). Essential eight. Retrieved July 8, 2025, from https://www.cyber.gov.au/resources-business-and-government/essential-cyber-security/essential-eight

Australian Local Government Association. (2024, July 2). 2024 National state of the assets report – Future proofing our communities. https://alga.com.au/2024-national-state-of-the-assets-report/

Boyd, S. (2024, November 15). *AI, drones and LiDAR as enablers of learning property*. Conference Presentation. 2024 CQUniversity Property Conference, Rockhampton, QLD, Australia.

Cook, K. (2024, July 3). *ALGA releases council asset management report*. Council Magazine. https://councilmagazine.com.au/alga-releases-council-asset-management-report/

Knight, W. (2021, March 31). *The foundations of AI are riddled with errors*. Wired. https://www.wired.com/story/foundations-ai-riddled-errors/wired.com

Marwala, T. (2024, July 18). *Never assume that the accuracy of artificial intelligence information equals the truth*. United Nations University. https://unu.edu/article/never-assume-accuracy-artificial-intelligence-information-equals-truth

Microsoft News Center. (2019, October 14). AI helps LG Valuations raise bar on rates, land tax accuracy. https://news.microsoft.com/en-au/features/ai-helps-lg-valuations-raise-bar-on-rates-land-tax-accuracy/

OpenAI. (2025, July 8). ChatGPT (o3 model) [Large language model]. https://chat.openai.com/

University of Maryland Libraries. (n.d.). Evaluate AI content for accuracy and bias [Library guide]. Retrieved July 8, 2025, from https://lib.guides.umd.edu/c.php?g=1340355&p=9880574

Value Australia. (n.d.). Valuations platform. Retrieved July 8, 2025, from https://value-australia.com.au/valuations-platform

Chapter 7

Airport runway inspections

Transitioning from conventional methods to drone-based solutions

Isara Khanjanasthiti, Armin Taklif and Samuel Gardner

Introduction

Runways and supporting infrastructure are essential assets for airport operations, as they facilitate takeoffs and landings for aircraft. As runways are consistently utilised, they are subject to ongoing wear and tear, thus requiring regular inspections to ensure the absence of structural damage and foreign objects and debris (FOD), both of which present safety risks to aircraft. Traditionally, these inspections are manually conducted by crewed vehicles and rely on the inspector's visual sense, aided by such tools as borescopes and cameras (Rodríguez et al., 2024). As such, these conventional visual inspections are susceptible to oversights.

Drone-based runway inspections have recently emerged as an alternative method and have garnered significant attention from airport operators and scholars worldwide (Ranjbar et al., 2023; Sappington et al., 2019; Wang & Sherry, 2022). Through the integration of digital technologies, this approach enables the rapid, consistent and accurate assessment of runway pavement conditions, thereby potentially enhancing the safety of aircraft operations. Drones also provide an additional perspective to assist in the physical inspections of runways. Nevertheless, many operational, legal and ethical implications are associated with using drones in an airport's heavily regulated airspace. Some of these include potential collisions between aircraft and drones, privacy concerns as passengers' movements may be recorded without consent and potential terrorism activities associated with the hacking of drones.

This chapter aims to evaluate the potential of using drones to monitor and manage runway conditions in Australian airports. To this end, the chapter comprises four sections. Firstly, the chapter conducts a literature review, examining the significance of runway inspections and comparing the advantages and disadvantages of conventional methods with those of drone-based solutions. Secondly, the chapter discusses the existing regulatory frameworks governing Australian airspace within and around airports, as well as their implications for the employment of drones for runway inspections. Thirdly, the chapter examines selected case studies of airports worldwide that have recently

DOI: 10.1201/9781003539254-9

employed drones for runway inspections. The utilisation of drones at these airports, including the associated challenges, opportunities and lessons, will be discussed. Lastly, the chapter concludes with recommendations for Australia's government agencies and airports, drawn from the analysis of both the regulations and the case studies.

The significance and evolution of runway inspections

Why do runway inspections matter?

The safety of runways is one of the critical aviation safety priorities specified by the International Civil Aviation Organisation (ICAO), a specialised agency of the United Nations. ICAO has created the Global Reporting Format (GRF), a standardised system for evaluating and reporting runway surface conditions that affect flight operations (Civil Aviation Safety Authority (CASA), 2024a). The GRF specifies standards for inspecting, assessing, reporting and using wet and contaminated runways under weather-related circumstances. It also highlights the importance of conducting regular inspections of runway conditions to ensure the safe operation of airports globally. Airport operators are only required to undertake runway surface condition inspection, assessment and reporting when aeroplane operations are scheduled or anticipated. The frequency of runway inspections varies significantly between airports. Major international airports typically conduct multiple inspections per day to ensure safety and operational efficiency. Smaller regional airports and aerodromes also perform inspections; however, their frequency—typically lower than that of major airports—depends on regulatory requirements, traffic volume and environmental conditions.

Conventional runway inspections

Traditionally, runway inspections, which have been vital to the ongoing operational safety of airports around the globe, are undertaken by trained airport staff in accordance with domestic and international reporting standards. This section provides an overview of the current inspection frameworks and requirements, followed by a discussion of the advantages and disadvantages of traditional runway inspections.

Australian airport operators are required to undertake and document regular inspections of runways and the surrounding infrastructure to ensure operational safety. Runway inspections are currently regulated by the standards prescribed by CASA, ICAO and the GRF. The frequency of surface inspections varies based on the size and volume of traffic, with the primary focus still being on identifying and documenting any defects in the runway surface or any FOD that might be present on runways (CASA, 2024b).

The Civil Aviation Safety Regulations 1998 (CASR), established under the Civil Aviation Act 1998, serve as Australia's principal framework for aviation safety. These regulations aim to ensure the safety of civil aviation operations and cover a broad range of areas, including aircraft operations, personnel licensing, air traffic services, aerodromes, safety management and compliance and enforcement. Under Part 139 of the CASR, airport operators must undertake a minimum of two scheduled serviceability inspections per week (CASA, 2023a). Inspections must be conducted at least 48 hours apart for airports with low traffic inflow or each day that there is a scheduled air transport movement. Scheduled inspections are primarily conducted from a vehicle, with multiple skilled operational officers required at larger airports. Meanwhile, unscheduled inspections are required after a major weather event, before low-visibility operations or following the identification of an incident or a significant hazard. Airside operation officers must conduct these inspections as soon as possible and document any findings in line with CASA's requirements for serviceability inspections to ensure the consistency of results across Australia.

Inspections of the aircraft manoeuvring area focus on detecting any maintenance or operational defects in the runway surface and pavement. These inspections cover areas used by Regular Public Transport (RPT) and general aviation (GA) operations, as well as associated technical lighting systems that support safe aircraft movement. Airside operations officers are also required to identify any FOD and various external risks and hazards that could risk the ongoing safety of operations. Other airport-specific risks, such as environmental and wildlife hazards, are also identified during this process.

Additional unscheduled inspections are required before the commencement of reduced or low-visibility operations, after a significant weather event, accident or incident or at the request of a pilot (CASA, 2023a). Airport operators are also required to complete an inspection if requested to do so by the aviation rescue firefighting services (ARFFS) provider or if there is reason to believe that a hazard to an aircraft might be present on a runway or a manoeuvring area. Runway inspections, the frequency of which is based on the extent of airport operations, involve reviewing the conditions of key infrastructure, including lighting systems and runway pavement.

Manual runway inspections must be conducted by appropriately trained personnel in accordance with regulatory standards. At controlled airports, these personnel must maintain consistent radio contact with Air Traffic Control (ATC). At non-controlled airports (where ATC is not present), runway inspection personnel must use the Common Traffic Advisory Frequency (CTAF) to coordinate with pilots and report any hazards identified during inspections.

The following are key advantages associated with the conventional method of runway inspections using airport personnel:

- **Enhancing understanding of environmental factors**: Traditional inspections provide an opportunity to develop a thorough understanding of the specific environmental factors that may influence the operations of runways and surrounding infrastructure. Airport environmental factors include soil, air, water, noise and chemical pollution. A thorough understanding of such factors is critical for airport operators in Australia to ensure ongoing compliance with various national legislative instruments, guidelines and standards (Australian Government, n.d.).

- **Fostering stakeholder relationships**: Manual runway inspections provide an opportunity to develop a dynamic team of stakeholders within major airports, which can assist in identifying risks between scheduled runway inspections (ICAO, 2015).

Disadvantages of traditional runway inspections include:

- **Exposure to human-based errors**: As personnel undertake inspections, they are susceptible to human-induced errors, meaning FOD and runway surface issues may be undetected. According to the Federal Aviation Administration (FAA, 1998), human error contributes to approximately 80% of airline accidents and incidents, thus highlighting the high likelihood of human-based errors.
- **Susceptibility to complacency**: Complacency is one of the most common causes of human error in airport settings (FAA, 2023). During a runway inspection, complacency can lead to a decline in the accuracy of the inspection. Utilising a 'tick and flick' checklist and relying on opinion-based input can have significant consequences for the safe operations of runways if FOD or other issues are not detected promptly. This makes complacency a significant ongoing risk for traditional runway inspections.
- **Vulnerability of regional airports**: Many regional airports operate ageing runways, necessitating more frequent inspections to ensure their operational safety. In Australia, the typical age of regional airports' runways is over 80 years old (ACIL Allen Consulting, 2016), which raises operational safety concerns for essential air services, such as the Royal Flying Doctor Service (RFDS), and other critical activities that regional and remote communities depend on. Regional airports also heavily rely on the limited number of trained inspectors based in the region and their ability to undertake comprehensive inspections on short notice. It also increases the risk of human error due to fatigue and poor decision-making.

As illustrated above, while conventional runway inspections have been consistently employed to maintain safe airport operations, they are subject to a range of challenges, entailing an alternative and more effective and efficient method—one that could employ drones.

Drone-based runway inspections

Drones are already extensively used throughout Australia, with such deployment projected to continue expanding. Nevertheless, the Australian Government (2024b) considers the drone sector in Australia to still be in the early stages of commercialisation and projects its application to grow rapidly across the following sectors:

- Emergency management and disaster recovery;
- E-commerce and delivery;
- Government and community services;
- Agriculture;
- Mining and resources;
- Defence; and
- Infrastructure inspections.

Some of the major applications of drones to date include emergency services, asset management, research, public safety and security, traffic monitoring and infrastructure inspection, pest detection in agricultural settings, air ambulance and medical supply delivery (Australian Government, 2024b). Drones are recognised by the Australian Government (2024b, p. 182) in its *Aviation White Paper: Towards 2050* as one of the 'emerging aviation technologies [that] will transform the use of airspace out to 2050'. Likewise, the Australian Government (2024a) has indicated its intention to promote the use of drone detection technology in its recently published *Australian National Aviation Safety Plan 2024–2027*. As such, it can be expected that drone usage will continue to increase throughout Australia, particularly in the context of airport infrastructure management.

The use of drones for runway inspections represents a significant advancement in the adoption of digital technologies within aviation infrastructure maintenance. Drones are increasingly integrated with advanced sensor systems such as LiDAR (Light Detection and Ranging), RGB cameras and thermal sensors, providing comprehensive data for evaluating runway conditions accurately and in real-time (Alonso et al., 2023; GlobeNewswire, 2019; Rodríguez et al., 2024). The imagery and data acquired by drones can be processed with computer vision and deep learning algorithms, such as convolutional neural networks (CNNs) and U-Net models, for precise identification and segmentation of pavement defects (Jiang et al., 2020; Maslan & Cicmanec, 2023; Parker et al., 2022).

Innovative systems, such as those trialled at Edmonton International Airport, have demonstrated how drones can support maintenance by capturing high-resolution images and 3D point clouds (GlobeNewswire, 2019). These images are analysed using artificial intelligence (AI) frameworks that enhance the detection of surface issues such as cracks, FOD and other anomalies (Papadopoulos & Gonzalez, 2021; Yang et al., 2024).

Key advantages of using drones for runway inspections include:

- **Improved safety and reduced human risk**: Drone inspections reduce the need for personnel to operate on active runways, minimising exposure to potential hazards (Maslan & Cicmanec, 2023; Tomić et al., 2020).

Table 7.2 Aviation legislation governing CASA and Australian airspace

Instrument	*Purpose(s)*
Tier 1 (Primary Legislation)	
Civil Aviation Act 1998	• Establishes the legal framework for regulating civil aviation safety in Australia • Creates CASA • Ensures compliance with international aviation safety standards
Airspace Act 2007	• Provides the framework for administering and regulating Australian airspace
Tier 2 (Delegated Legislation)	
Civil Aviation Regulations 1988	• Prescribes safety standards for several aspects of aviation (e.g., flight operations and licensing)
Civil Aviation Safety Regulations 1998	• Expand the Civil Aviation Regulations with more detailed and specific safety requirements
Airspace Regulations 2007	• Detail the classification and management of Australian airspace
Tier 3 (Regulatory Requirements and Standards)	
Manuals of Standards	• Provide detailed technical standards and requirements complementing the Civil Aviation Safety Regulations 1998
Civil Aviation Orders	• Prescribe mandatory operational, airworthiness and safety requirements to ensure consistent and safe aviation practices
Other legislative instruments	These include, but are not limited to: • Australian Technical Standard Orders; • Airworthiness Directives; • Authorisations; • Approvals; • Designations; • Directions; • Exemptions; and • Revocation notices.

(Table 1.2 Is Created by the Authors, With Information Sourced from Casa (2023b))

Australian airspace is divided broadly into two major types: controlled and uncontrolled. In controlled airspace, all air traffic is actively monitored and managed by ATCs, and an aircraft must receive a clearance from an ATC before entering the airspace. In uncontrolled airspace, there is no ATC supervision, and no clearance is needed for an aircraft to enter the airspace. The majority of light aircraft and helicopters operate in uncontrolled airspace (Airservices Australia, n.d.). There are five different classes of Australian airspace, namely A, C, D, E and G. Classes A, C, D and E are 'controlled' airspace, whereas Class G is 'non-controlled' airspace (CASA, n.d.-b). Figure 7.1 visually illustrates the different classes of Australian airspace, while Table 7.3 summarises their characteristics, including altitude, description and permitted flights.

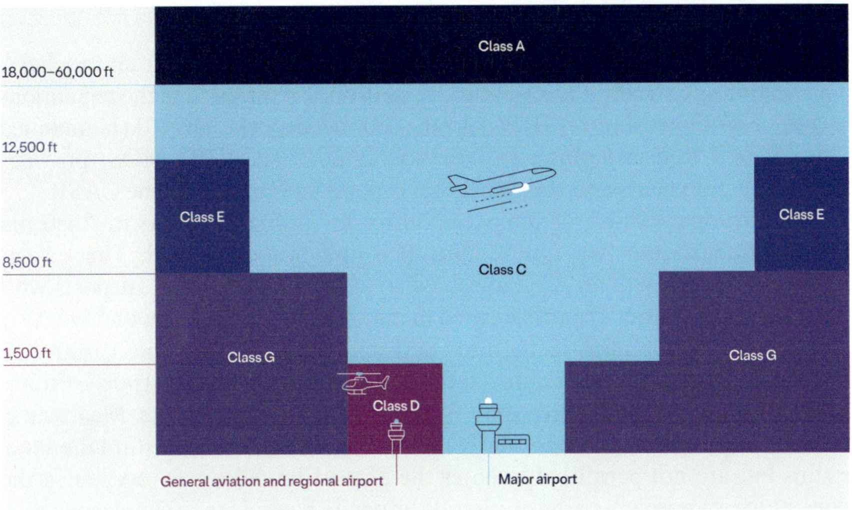

Figure 7.1 Australian airspace architecture and its different classes (Figure 7.1 is reproduced in its original form with permission from Airservices Australia, which retains full copyright. © Airservices Australia 2025.

Table 7.3 Australian airspace classes[a]

Class	*Altitude*	*Description*	*Permitted flights*
A	18,000 to 60,000 feet	High-altitude en route *controlled* airspace, primarily used by commercial and passenger aircraft	Instrument Flight Rules (IFR) only
C	0 to 17,999 feet	*Controlled* airspace surrounding major airports	IFR and Visual Flight Rules (VFR)
D	0 to 1,500 feet	*Controlled* airspace surrounding general aviation and regional airports that are equipped with a control tower	IFR and VFR
E	8,500 to 12,499 feet	Mid-altitude en route *controlled* airspace	IFR and VFR
G	0 to 8,499 feet	*Uncontrolled* airspace outside the controlled airspace of an airport	IFR and VFR

Table 7.3 is created by the authors, with information sourced from Airservices Australia (n.d.)

[a] The class of airspace at towered airports can change, subject to the time of day. IFR (Instrument Flight Rules) flights refer to flights where pilots rely on instruments for navigation and control, rather than relying on visual cues. VFR (Visual Flight Rules) flights are flights where pilots navigate using visual references outside the aircraft, such as landmarks and the horizon.

Legal implications of airspace regulations

Part 101 of the CASR outlines the rules for 'Unmanned aircraft and rockets'. This section specifically addresses the use of drones, referred to in the regulations as 'remotely piloted aircraft' (RPA), in and around airports. Part 101 (Unmanned Aircraft and Rockets) Manual of Standards 2019 (Part 101 MOS) provides detailed requirements and standards that expand on Part 101 of the CASR.

From the perspective of airspace control for drones, airports in Australia are categorised into two types: controlled and non-controlled. The former refers to airports with an ATC tower, whereas the latter refers to airports with no ATC tower. Busier airports located in major cities typically have active ATC services due to the need to monitor and manage aircraft traffic. Controlled airports have stricter regulations on the use of drones. In such airports, drones over 250 g are not permitted to fly within 5.5 km of the airports. Meanwhile, drones weighing 250 g or less can fly up to 45 metres (150 feet) within the same radius but are not permitted to enter the airport boundary, fly over or in the departure or approach paths of aircraft or create hazards to other aircraft during their take-off or landing. In non-controlled airports, which are the majority of Australian airports, drones can be flown within 5.5 km of the airport. Nevertheless, when drone operators become aware of an aircraft, their drone must not be launched, and any existing drones in the air must be landed as soon as possible. Figures 7.2 and 7.3 outline drone operation restrictions applicable to controlled and non-controlled airports.

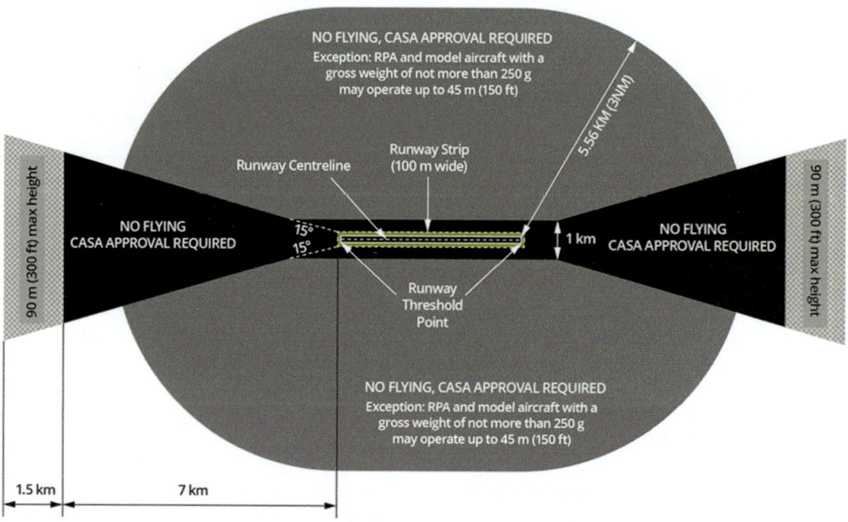

Figure 7.2 Drone operation restrictions for controlled airports in Australia (CASA, 2024c) (Figure 7.2 is reproduced in its original form under the CC BY 4.0 License (CASA, 2025; Creative Commons, n.d.)).

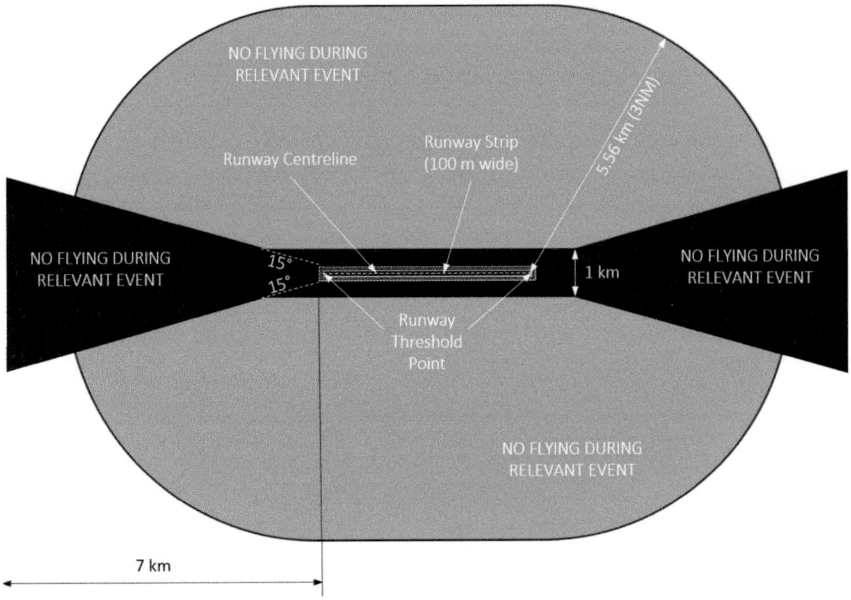

Figure 7.3 Drone operation restrictions for non-controlled airports in Australia (CASA, 2024g) (Figure 7.3 is reproduced in its original form under the CC BY 4.0 License (CASA, 2025; Creative Commons, n.d.)). A 'relevant event' is defined by CASA (2024, p. 11) as 'When a manned aircraft is within relevant airspace, including when the aircraft is in the course of approaching, landing at, taking off from, or manoeuvring on the movement area of, the [airport].'

While the restrictions related to 'no-fly zones' above apply to all recreational drone users, commercial drone operators with an RPA Operators Certificate (ReOC) issued by CASA can apply for permissions to operate drones beyond these restrictions. ReOC holders may seek approval from CASA to fly their drones within 5.5 km of controlled airports. Depending on the airspace classification, approvals may also be required from Airservices Australia, which has jurisdiction over civilian airspace, or the Department of Defence, which controls military airspace (CASA, 2024g). They are also permitted to employ drones heavier than two kilograms (CASA, 2024d). Importantly, heavier drones enable the use of more advanced equipment and cameras, which can assist in runway inspection procedures. ReOC holders are also permitted to operate their drones at night, which is critical for monitoring airport runways with night flights.

Therefore, ReOCs are essential for drone operations for runway inspections at controlled airports, which are located in major cities. Meanwhile, drone operations in non-controlled airports—or airports situated in regional, rural or remote Australia—can be conducted with significantly fewer restrictions. Applications for ReOCs may 'take up to 70 business days', or approximately 3 months, depending on the complexity of the drone operation (CASA, 2024d).

In 2021, CASA conducted a review of the CASR and the Part 101 MOS, leading to proposed amendments to both legislative instruments. This proposal was advertised for public feedback between December 2021 and February 2022 (CASA, 2022). As of October 2024, Part 101 MOS is undergoing several changes, which will affect ReOC holders and provide 'clarified requirements for RPA … operating in the no-fly zone of controlled [airports and] near non-controlled [airports]' (CASA, 2024f). The Australian Government (2024b, p. 182) has also recently expressed its intention to consider introducing and mandating new legislation by 2030 'to protect Australian communities, infrastructure and businesses from security risks of drones'. As such, the already extensive regulations around Australian airspace and drones will potentially become more complex.

The comprehensive legislation around the operation of drones reflects Australia's history as the first country to regulate their usage (Library of Congress, 2016). Meanwhile, Australian airspace legislation reflects the rigorous global aviation standards prescribed by ICAO. Together, these regulations and standards can help ensure the safety of aircraft operations during runway inspections conducted by drones. However, they may also be perceived as significant legislative and time barriers to such inspections due to the need for drone operators to not only obtain ReOCs—a potentially lengthy and complex process—but also stay current with the constantly amended legislation. Nevertheless, operators of drones for runway inspections may benefit from the following intentions to enable drones as 'new aviation technologies', which are outlined by the Australian Government (2024b, p. 181):

- Implementation of a new system to facilitate electronic communication between air traffic management and drones through the *Flight Information Management System*; and
- Introduction of *Remote ID* requirements for drones as part of the new legislation to be introduced and mandated by 2030.

Ethical challenges

The employment of drones for runway inspections introduces a suite of ethical challenges, primarily centred on community concern, privacy, cybersecurity and public trust, which are further discussed below.

Community and safety concerns

Large-scale uncoordinated or unchecked deployment of drones can potentially lead to significant risks in terms of 'community concern around noise, safety and security' (Australian Government, 2024b, p. 182). For runway inspections, drones are typically deployed at lower altitudes than aircraft, which can exacerbate the risk of collision with aircraft. As such, consistent and ongoing coordination of drone and aircraft operations, including regular correspondence between the ATC and runway inspection departments, is necessary to alleviate community and safety concerns.

Privacy issues

Drones equipped with high-resolution cameras and advanced sensors can inadvertently capture sensitive data, including personal activities of passengers and airport staff. Cavoukian (2012, p. 16) highlights the potential for 'surveillance by design', necessitating stringent adherence to privacy principles such as 'Privacy by Design' to embed safeguards into drone operations. For airports, this approach involves implementing strict protocols for data collection, storage and sharing, ensuring compliance with relevant privacy legislation. Furthermore, transparency in how drone data is used can also help mitigate public apprehension and uphold trust.

Cybersecurity risks

The absence of robust cybersecurity standards for drones in Australia poses a significant ethical challenge. The Cyber Security Cooperative Research Centre (CSCRC) and Omni (2024) note that drones are vulnerable to hacking, unauthorised access and data breaches due to their reliance on internet connectivity for regular updates and communication. The potential for cyberattacks extends beyond operational disruptions to national security risks in critical infrastructure settings such as airports. This highlights the ethical imperative to prioritise robust encryption, multi-layered authentication and proactive cybersecurity frameworks to safeguard airport drone operations.

Balancing innovation and public trust

The ethical implications of drone deployment also encompass community concerns about excessive surveillance and workplace dynamics. Drones can evoke fears of over-monitoring amongst employees, leading to tensions that may hinder collaboration and morale (Al-Dosari & Fetais, 2023; Khalid et al., 2021). Public engagement and transparent communication are crucial for addressing these concerns and aligning drone usage with societal values and expectations. Khalid et al. (2021) emphasise the importance of balancing technological innovation with ethical responsibility to establish and maintain public trust.

Recommendations

Addressing these ethical challenges is vital to ensuring the responsible deployment of drones within Australia's heavily regulated airport environments.

To mitigate these challenges, Australian airports must adopt comprehensive frameworks that address both privacy and security. Drawing from Mekdad et al. (2023), proactive measures such as real-time intrusion detection systems, advanced encryption methods and clear consent protocols for data capture are essential. Additionally, embedding 'Privacy by Design' principles into drone systems and engaging stakeholders both internally and externally in collaborative discussions can foster a culture of trust and accountability. By proactively addressing these ethical considerations, drone technology can be leveraged to enhance runway safety and operational efficiency without compromising privacy, security or public confidence.

Table 7.4 provides a summary of the legal and ethical challenges for drone-based runway inspections.

Table 7.4 Summary of legal and ethical challenges for drone-based runway inspections

Australian airspace regulations

- CASA and Airservices Australia regulate Australian airspace in line with ICAO standards.
- Airspace is classified into controlled (ATC-managed) and uncontrolled (no ATC).
- Key legislation includes the Civil Aviation Act 1988 and the Airspace Act 2007.
- Airspace is divided into five classes (A, C, D, E, and G) with specific rules for permitted flights.

Legal implications of airspace regulations

- Part 101 of the CASR and Part 101 MOS govern drone operations near airports.
- In controlled airports, drones over 250 g are banned within 5.5 km; lighter drones are restricted in altitude and must avoid critical flight paths.
- In non-controlled airports, drones can fly within 5.5 km but must yield to crewed aircraft.
- ReOC holders can apply for exemptions, use drones over 2 kg and operate at night.
- Proposed updates to legislation aim to introduce Remote ID and better ATC-digital integration by 2030, though ReOC application processes remain lengthy.

Ethical challenges

- Ethical issues relate to safety, privacy, cybersecurity and public trust.
- Low-altitude drone use raises risks of collision and community anxiety; coordination with ATC is critical.
- Drones can inadvertently record sensitive personal data; compliance with privacy legislation is essential.
- Drones are vulnerable to hacking and data breaches; cybersecurity protections are needed.
- Concerns about over-surveillance and workplace impacts may reduce public and staff trust; transparency and engagement are key.
- Ethical operations require intrusion detection, data safeguards and stakeholder collaboration to balance innovation with accountability.

Following an examination of the legal and ethical challenges associated with employing drones for airport runway inspections, the chapter next analyses three case studies of airports that have recently adopted drone technology for this purpose, highlighting key insights from their experiences.

Case studies

Grosse Ile Municipal Airport

Case study overview

Grosse Ile Municipal Airport (ONZ) is located in Wayne County, Michigan, on Grosse Ile, approximately 3.2 km south of Grosse Ile Township. It serves as a public-use reliever airport and features two primary runways: Runway 4/22 and Runway 17/35. Runway 4/22, measuring 1,477 m by 30.5 m, is surfaced with asphalt and has an overall pavement condition rated as ‘Good’. In contrast, Runway 17/35, a 1,348.5 m by 22.86 m Portland cement concrete (PCC) runway, has extensive peeling and cracking, leading to a ‘Poor’ rating. This runway has been subject to various issues, such as alkali-silica reaction, corner breaks, extensive patching, faulting, D-cracking, joint seal damage and shattered slabs (Sourav et al., 2023a; Sourav et al., 2023b).

Drone deployment for runway inspections

To enhance the airport’s Pavement Management Programme (PMP), drones were deployed to evaluate runway conditions comprehensively. A combination of drones equipped with high-resolution RGB optical sensors and stereo thermal cameras was used for this purpose. The data collection took place at varying altitudes to balance image resolution and coverage area. The Bergen Hexacopter and DJI Mavic 2 Pro were amongst the platforms used, capturing high-quality orthophotos and thermal images. RGB orthophotos offered detailed visual assessments of cracks, while Digital Elevation Models (DEMs) provided 3D representations to identify elevation-based distress such as faulting (Sourav et al., 2023a; Sourav et al., 2023b).

Challenges, opportunities and lessons

One of the primary challenges was weather dependency, which influenced image clarity and sensor effectiveness. Optimal sunlight and minimal wind conditions were critical for high-quality data. Another challenge was the limited automated data collection support at lower altitudes, necessitating manual adjustments and careful flight planning.

The deployment of drones provided significant opportunities, including faster data acquisition compared to traditional inspections and enhanced personnel safety. Additionally, drones enabled comprehensive coverage of the runways, reducing the risk of missing critical pavement distress.

This case study emphasises the importance of integrating high-resolution orthophotos with DEMs for thorough distress analysis. The use of multi-sensor approaches allowed for more accurate detection of both surface and subsurface issues, confirming that drone-based inspections could effectively complement traditional PMP methods (Sourav et al., 2023a; Sourav et al., 2023b).

Custer Airport

Case study overview

Custer Airport (TTF) is a public-use airport located approximately 3.2 km northwest of Monroe, Michigan. The airport features a single runway, Runway 03/21, measuring 1,523 m by 30 m, surfaced with grooved asphalt and rated in good condition. The primary taxiway, Taxiway A, is partially surfaced with asphalt concrete and grooved asphalt. Previous inspections indicated that Runway 03/21 had a Pavement Condition Index (PCI) score of 82 in 2019, which decreased to 74 by 2021 due to various distress types, including longitudinal and transverse (L&T) cracks, swelling and weathering (Sourav et al., 2023a; Sourav et al., 2023b).

Drone deployment for runway inspections

Drone platforms such as the mdMapper1000+ and Mavic 2 Enterprise Advanced were deployed to perform detailed inspections. These drones were equipped with advanced sensors, including a 42.4-megapixel RGB camera and high-resolution thermal cameras. Data collection was performed at multiple altitudes to create orthophotos, DEMs and thermal images. The orthophotos provided a detailed visual assessment of surface cracks and weathering, while the thermal images were particularly effective for identifying subsurface anomalies and potential moisture retention areas (Sourav et al., 2023a; Sourav et al., 2023b).

Challenges, opportunities and lessons

A significant challenge was the precise placement and alignment of Ground Control Points (GCPs) to ensure accurate overlay of geospatial data. This required meticulous planning and manual setup, resulting in increased preparation time. Nevertheless, the opportunity for more extensive and faster coverage compared to traditional inspections was notable. The use of drones significantly reduced the time required for inspections while maintaining high accuracy in detecting runway issues. Additionally, the thermal imaging feature of the drones added value by revealing underlying structural problems that might not be visible through standard visual inspections.

The primary lesson from the case study is the effectiveness of combining thermal and optical imaging to provide a more comprehensive analysis.

The integration of drone data with Geographic Information System (GIS) platforms enabled more accurate visualisation and tracking of pavement conditions over time (Sourav et al., 2023a; Sourav et al., 2023b).

Northern Canada's Remote Airports

Case study overview

Remote airports in Northern Canada play a crucial role in connecting isolated communities and facilitating essential services. These airports predominantly feature gravel runways, with 94 out of Canada's 117 remote airports relying on unsurfaced gravel as their primary landing strips. These runways are vital for transportation in the vast, remote landscapes of Northern Canada, where road access is often not feasible. Ensuring the safety and structural integrity of these gravel runways is essential for maintaining the reliability of air transport in the region (Yang et al., 2024).

Drone deployment for runway inspections

To address the unique challenges associated with monitoring and maintaining gravel runways, a novel system integrating drones and computer vision technologies was developed. The system was designed to automatically detect runway defects such as surface water pooling, vegetation encroachment and surface irregularities. Drones equipped with high-resolution cameras captured detailed images, which were then processed using deep learning algorithms, including a vision transformer model and image filtering techniques, to identify and evaluate the condition of the runways (Yang et al., 2024).

Challenges, opportunities and lessons

Challenges included operating in harsh climates, which posed difficulties for both data collection and equipment durability. The remoteness of these airports also added logistical hurdles, such as the need for specialised teams and equipment transportation. Despite these challenges, the deployment of drones offered significant opportunities. It provided a cost-effective and efficient alternative to traditional, labour-intensive manual inspections, which are often infrequent due to high operational costs and limited accessibility of the airports. The use of drones also enabled more frequent data collection and access to areas that would be difficult to inspect manually. The automated system generated comprehensive reports that included numerical data and visual representations, aiding in timely maintenance decisions.

Insights that can be drawn from the case studies include the need for robust training datasets for the deep learning algorithms, as detecting defects such as water pooling and vegetation proved challenging without sufficient labelled data examples. The system's application demonstrated that drone-based inspections could significantly enhance the safety and maintenance practices of remote airfields in Canada (Yang et al., 2024).

Table 7.5 presents a summary of the key findings from the three case studies.

Table 7.5 Summary of key findings from case studies

Grosse Ile Municipal Airport (ONZ)	
Overview	• Located in Michigan, ONZ has two runways: Runway 4/22 (asphalt, rated 'Good') and Runway 17/35 (PCC, rated 'Poor'). • Runway 17/35 suffers from alkali-silica reaction, D-cracking, corner breaks, joint seal damage, faulting and shattered slabs.
Drone deployment	• Drones with RGB and stereo thermal cameras were deployed for comprehensive inspections. • Orthophotos enabled detailed surface assessments, while Digital Elevation Models captured elevation-related distress.
Challenges, opportunities and lessons	• Weather conditions affected image clarity and sensor accuracy; low-altitude data collection lacked automation. • Drones enabled quicker inspections and improved safety compared to traditional methods, while reducing the risk of missed defects. • Multi-sensor data integration proved vital for accurate surface and subsurface analysis, supporting improvements to the airport's Pavement Management Programme.
Custer Airport (TTF)	
Overview	• Situated near Monroe, Michigan, TTF has one grooved asphalt runway (Runway 03/21), which has been deteriorating due to cracks, swelling and weathering
Drone deployment	• Drones with high-res RGB and thermal sensors were used. • Orthophotos and thermal imaging supported visual crack detection and identification of subsurface anomalies (e.g. moisture).
Challenges, opportunities and lessons	• To ensure accurate overlay of geospatial data, precise placement of Ground Control Points required meticulous manual setup, thus increasing preparation time. • Drones enabled faster, broader inspections with high precision. • Thermal and optical imaging integration enhanced subsurface detection. GIS integration improved data tracking over time.

(*Continued*)

Table 7.5 (Continued)

Northern Canada's remote airports	
Overview	• Most remote airports in Northern Canada rely on gravel runways, which are vital for isolated communities.
Drone deployment	• Drones with high-resolution cameras paired with deep learning models (e.g. vision transformers) were used to detect defects such as pooling, vegetation and surface irregularities.
Challenges, opportunities and lessons	• Harsh climates and remoteness posed operational and logistical challenges. • Drones enabled cost-effective, frequent inspections and access to difficult-to-reach areas. • Automated defect detection improved maintenance response and safety. • Limitations in training datasets hindered some detections (e.g. water pooling), highlighting the need for more labelled data.

Having reviewed the case studies of drone applications for airport runway inspections, the next section of this chapter presents its conclusion. It synthesises the key discussion points and offers recommendations to facilitate the effective use of drones in runway inspections.

Conclusion

Towards drone-based solutions

The safety of aircraft operations at airports necessitates ongoing inspections of runways. Although runway inspections are conventionally conducted by crewed vehicles, such methods are prone to human errors and pose safety concerns for airport personnel. Emerging technologies, particularly drones, hold significant potential in improving both the efficiency and the effectiveness of runway inspections.

The case studies of Grosse Ile Municipal Airport, Custer Airport, and remote airports in Northern Canada highlight the effectiveness and challenges of using drones for runway inspections in various airport contexts. Each deployment showed that drones significantly improve inspection efficiency, data accuracy and safety compared to traditional methods. For instance, the drones' combination of thermal and optical imaging allowed for the precise detection of surface and subsurface issues, while GIS integration enabled detailed tracking of runway conditions. However, challenges such as weather dependency, accurate placement of Ground Control Points and harsh operating environments in remote locations underscore the need for further technological advancements and adaptations to ensure optimal performance.

From a legislative standpoint, airspace regulations, while essential for aviation safety, can significantly restrict drone operations. The case studies highlight the importance of evolving airspace laws to better support drone-enabled runway inspections. This is particularly relevant in Australia, where regulatory frameworks for airport-adjacent airspace remain extensive and frequently updated, often limiting the practical use of drones.

With a conducive regulatory environment and targeted investment in digital tools, drones can become a cornerstone of smarter, safer and more sustainable runway inspection strategies across Australia and beyond. The following recommendations outline strategies to facilitate drone use for runway inspections, present key questions for readers, and provide relevant resources related to this chapter.

Recommendations for agencies and readers

In both current and future revisions of airspace legislation, CASA should aim to facilitate—rather than hinder—the use of drones for airport infrastructure surveillance. It should also develop specific protocols to enable safer and more efficient drone operations for runway inspections. Furthermore, airport operators should explore integrating emerging digital technologies and software into drones, such as deep learning and AI-driven analytics, which could further enhance the effectiveness of runway inspections, particularly in real-time analysis and defect identification. These suggestions are grounded in the case studies, which underscore the value of regulatory reforms and advanced digital solutions in realising the full potential of drone inspections in diverse airport environments.

The following are four key questions for further discussion and reflection by the readers:

1 Should drones completely replace personnel in the process of airport runway inspections? Why or why not?
2 How can government agencies facilitate, rather than restrict, the use of drones for runway inspections?
3 How can drones be employed on a large scale for runway inspections safely and equitably?
4 Can drone technology help regional airports inspect their ageing runways more efficiently—and if so, how?

The following are recommended resources for further reading

Australian Government. (2024). *Aviation White Paper—Towards 2050*. Australian Government. https://bit.ly/3CESWF3

Cavoukian, A. (2012). *Privacy and Drones: Unmanned Aerial Vehicles*. Information and Privacy Commissioner of Ontario. https://bit.ly/4i4qInu

CSCRC, & Omni. (2024). *Flight Critical: Drones, Cyber Security and Critical Infrastructure*. Australian Government. https://bit.ly/3ZkLacf

Papadopoulos, E., & Gonzalez, L. F. (2021, 6-13 March). UAV and AI Application for Runway Foreign Object Debris (FOD) Detection. *2021 IEEE Aerospace Conference, Online*.

Sourav, M. A. A., Ceylan, H., Brooks, C., Peshkin, D., Kim, S., Dobson, R., Cook, C., Mahedi, M., & Jenkins, A. (2023). *Small Unmanned Aircraft System for Pavement Inspection*. Federal Aviation Administration.

References

ACIL Allen Consulting. (2016). *Regional Airport Infrastructure Study: Economic Contribution and Challenges of Regional Airports in Australia*. Australian Government.

Airservices Australia. (n.d.). *How Airspace is Managed*. Airservices Australia. https://bit.ly/4f9NKHd

Al-Dosari, K., & Fetais, N. (2023). A New Shift in Implementing Unmanned Aerial Vehicles (UAVs) in the Safety and Security of Smart Cities: A Systematic Literature Review. *Safety*, *9*(3), 64.

Alonso, P., Iñiguez de Gordoa, J. A., Ortega, J. D., García, S., Iriarte, F. J., & Nieto, M. (2023). Automatic UAV-Based Airport Pavement Inspection Using Mixed Real and Virtual Scenarios. In *Proceedings of the SPIE Fifteenth International Conference on Machine Vision, (Art. 1270118)*, pp. 361–372. https://doi.org/10.1117/12.2679734

Australian Government. (2023). *Western Sydney International (Nancy-Bird Walton) Airport — Airspace and Flight Path Design: Draft Environmental Impact Statement*. Australian Government.

Australian Government. (2024a). *The Australian National Aviation Safety Plan 2024–2027*. Australian Government. https://bit.ly/4fHgSGj

Australian Government. (2024b). *Aviation White Paper—Towards 2050*. Australian Government. https://bit.ly/3CESWF3

Australian Government. (n.d.). Airport Environmental Management. https://bit.ly/3VadCLt

CASA. (2022). Proposed amendments to Part 101 CASR and MOS - Unmanned Aircraft and Rockets – (PP 2107US). https://bit.ly/3YEliHW

CASA. (2023a). *Advisory Circular AC 139.C-03 v1.1: Serviceability Inspections*. CASA.

CASA. (2023b). CASA's Regulatory Framework. https://bit.ly/4hmCVUa

CASA. (2024a). *AC 91-32 and AC 139-22 v1.0 Global Reporting Format – Runway Surface Condition*. CASA.

CASA. (2024b). Assessing and Reporting Runway Surface Conditions. https://bit.ly/3AWlQ2U

CASA. (2024c). Flying Near Airports. https://bit.ly/4e4LxMe

CASA. (2024d). *Get your ReOC*. CASA. https://bit.ly/3NY2XQh

CASA. (2024e). *How Airspace Is Regulated*. Australian Government. https://bit.ly/48iEuyr

CASA. (2024f). Part 101 of CASR Unmanned Aircraft and Rockets Changes. https://bit.ly/4e3Jx6T

CASA. (2024g). *Remotely Piloted Aircraft Systems - Licensing and Operations*. CASA. https://bit.ly/3YAzoKB

CASA. (2025). *Copyright*. CASA. Retrieved 26 April from https://bit.ly/42MkEcd
CASA. (n.d.-a). Airspace Regulation. https://bit.ly/4hgymdP
CASA. (n.d.-b). *Australian Airspace Structure*. Australian Government. https://bit.ly/3UgDdSo
Civil Aviation Safety Regulations, § Part 101 (1998). https://bit.ly/4hgyuu0
Creative Commons. (n.d.). CC BY 4.0: Attribution 4.0 International – Deed. Retrieved 26 April from https://bit.ly/3Gmuqu7
CSCRC, & Omni. (2024). *Flight Critical: Drones, Cyber Security and Critical Infrastructure*. Australian Government. https://bit.ly/3ZkLacf
FAA. (1998). *Human Factors in Aviation Maintenance and Inspection: Strategic Plan*. FAA. https://bit.ly/4eMBig2
FAA. (2023). Human Factors in Aviation Maintenance. https://bit.ly/498IfqC
GlobeNewswire, I. (2019). EIA Canadian Leader in Using Drones for Safety: AERIUM Partnership Saves Time, Money on Runway Inspections. *NASDAQ OMX's News Release Distribution Channel*. https://www.globenewswire.com/news-release/2019/10/22/1933499/0/en/EIA-Canadian-leader-in-using-drones-for-safety.html
ICAO. (2015). *Runway Safety Team Handbook*. ICAO. https://bit.ly/3Z4UtM1
Jiang, L., Xie, Y., & Ren, T. (2020). A Deep Neural Networks Approach for Pixel-Level Runway Pavement Crack Segmentation Using Drone-Captured Images. *arXiv preprint arXiv:2001.03257*.
Khalid, M., Namian, M., & Behm, M. (2021). Safety Implications of Using UAVs in Construction: An Ethical Perspective. *Joint CIB W099 & W123 International Conference 2021*, Glasgow, Scotland.
Kim, S., Paes, D., Lee, K., Irizarry, J., & Johnson, E. N. (2019). UAS-Based Airport Maintenance Inspections: Lessons Learned from Pilot Study Implementation. *Computing in Civil Engineering 2019*, Atlanta, Georgia.
Library of Congress. (2016). *Regulation of Drones: Australia, Canada, People's Republic of China, France, Germany, Israel, Japan, New Zealand, Poland, South Africa, Sweden, Ukraine, United Kingdom, European Union*. Federation of American Scientists. https://irp.fas.org/congress/2016_rpt/lloc-drones.pdf
Maslan, J., & Cicmanec, L. (2023). A System for the Automatic Detection and Evaluation of the Runway Surface Cracks Obtained by Unmanned Aerial Vehicle Imagery Using Deep Convolutional Neural Networks. *Applied Sciences*, *13*(10), 6000.
Mekdad, Y., Aris, A., Babun, L., El Fergougui, A., Conti, M., Lazzeretti, R., & Uluagac, A. S. (2023). A Survey on Security and Privacy Issues of UAVs. *Computer Networks*, *224*, 109626.
Papadopoulos, E., & Gonzalez, L. F. (2021, 6–13 March). UAV and AI Application for Runway Foreign Object Debris (FOD) Detection. *2021 IEEE Aerospace Conference, Online*.
Parker, A., Gonzalez, F., & Trotter, P. (2022, 5-12 March). Live Detection of Foreign Object Debris on Runways Detection using Drones and AI. *2022 IEEE Aerospace Conference (AERO)*,
Ranjbar, H., Forsythe, P., Fini, A. A. F., Maghrebi, M., & Waller, T. S. (2023). Addressing Practical Challenge of Using Autopilot Drone for Asphalt Surface Monitoring: Road Detection, Segmentation, and Following. *Results in Engineering*, *18*, 101130. https://doi.org/10.1016/j.rineng.2023.101130
Rodríguez, D. A., Lozano Tafur, C., Melo Daza, P. F., Villalba Vidales, J. A., & Daza Rincón, J. C. (2024). Inspection of Aircrafts and Airports Using UAS: A Review. *Results in Engineering*, *22*, 102330. https://doi.org/10.1016/j.rineng.2024.102330

Sappington, R. N., Acosta, G. A., Hassanalian, M., Lee, K., & Morelli, R. (2019). *Drone Stations in Airports for Runway and Airplane Inspection Using Image Processing Techniques AIAA Aviation 2019 Forum*, Reston, Virginia.

Sourav, M. A. A., Ceylan, H., Brooks, C., Peshkin, D., Kim, S., Dobson, R., Cook, C., Mahedi, M., & Jenkins, A. (2023a). *Small Unmanned Aircraft System for Pavement Inspection*. Federal Aviation Administration.

Sourav, M. A. A., Mahedi, M., Ceylan, H., Kim, S., Brooks, C., Peshkin, D., Dobson, R., & Brynick, M. (2023b). Evaluation of Small Uncrewed Aircraft Systems Data in Airfield Pavement Crack Detection and Rating. *Transportation research record*, *2677*(1), 653–668. https://doi.org/10.1177/03611981221101030

Tomić, L., Cokorilo, O., & Macura, D. (2020). Runway Pavement Inspections Using Drones – Safety Issues and Associated Risks. *International Journal for Traffic and Transport Engineering*, *10*(3), 278–285.

Wang, C., & Sherry, L. (2022). Design of a Local Area Integrated Drone, Aircraft, Vehicle, and Asset Management System (LAIDAVAMS) *2022 Integrated Communication, Navigation and Surveillance Conference*,

Yang, Z., Nashik, S., Huang, C., Aibin, M., & Coria, L. (2024). Next-Gen Remote Airport Maintenance: UAV-Guided Inspection and Maintenance Using Computer Vision. *Drones*, *8*(6), 225. https://www.mdpi.com/2504-446X/8/6/225

Chapter 8

Generative AI prompt engineering basics for real estate appraisers

Keith Wolf and Johari Hussein Nassor Amar

Introduction

Recent developments in generative artificial intelligence (Gen AI) have inspired many industries to integrate Gen AI into everyday practices to improve efficiency and speed. In real estate, professionals are adopting Gen AI for tasks such as data analysis, efficiency and marketing, with appraisers increasingly exploring its potential to enhance valuation processes. Used responsibly, Gen AI can improve business operations and data analytics. However, used irresponsibly, it poses ethical, legal and socioeconomic risks, particularly in sectors where accuracy and impartiality are critical.

In real estate appraisal, three areas present the greatest regulatory and financial recourse liability when using Gen AI: (1) ethical and legal compliance, including the Fair Housing Act, a landmark United States law enacted in 1968 that prohibits discrimination in the sale, rental and financing of housing based on race, religion, national origin and other protected characteristics (Fair Housing Institute, 2020); (2) appraisal review processes where Gen AI may be used to evaluate appraisal reports; and (3) time adjustments, where Gen AI can assist in analysing market trends. Appraisers must be aware of these risks, as any misinformation or bias can compromise the integrity of valuation reports and expose practitioners to significant liability. This chapter provides a practical examination of Gen AI applications within the real estate valuation field. It focuses on three case studies that illustrate Gen AI's role in detecting biased language, supporting appraisal review functions and deriving market-based time adjustments.

Specific skills are needed in the use of Gen AI. To use Gen AI effectively, the user should have the basic skills necessary to ask Gen AI the correct questions to mitigate hallucinations. Gen AI needs specific prompts for the tasks they are asked to complete. Gen AI can only do what the user asks; it cannot read the user's mind. Users need to understand coding methods and have basic data analysis skills. Additionally, and most importantly, the user needs to be a domain expert in the topic. In this chapter, the area of expertise is real estate valuation. This chapter also explores best practices for prompt engineering and

DOI: 10.1201/9781003539254-10

discusses the skills appraisers need to use AI responsibly, highlighting the importance of domain expertise, critical evaluation and technical understanding when incorporating AI tools into professional practice.

This chapter outlines a clear process users can follow to reduce the risk of hallucinations in generative AI. Providing context, including background information about the problem being addressed, is a key step. It draws on real-life data and an appraisal report prepared by the author to show how generative AI, using Julius AI as an example, may produce incorrect or misleading outputs. Julius AI is an ensemble model that combines several platforms focused on analytics and research. Case studies are used to demonstrate practical prompt techniques and examine the model's responses.

Understanding generative artificial intelligence (Gen AI)

Generative artificial intelligence (Gen AI) is a subset of artificial intelligence focused on creating new content such as text, images and data outputs based on large volumes of training information (see Figure 8.1). In real estate practice, Gen AI offers the potential to accelerate tasks such as data analysis, report writing and market research. However, while the technology can enhance efficiency, it does not independently verify the accuracy or appropriateness of its outputs.

Gen AI models rely on large language models (LLMs) and machine learning (ML) algorithms that identify patterns in vast datasets, which may be structured (labelled fields such as addresses, prices and property types) or unstructured (free-form text, images or documents) Kalota, 2024). These

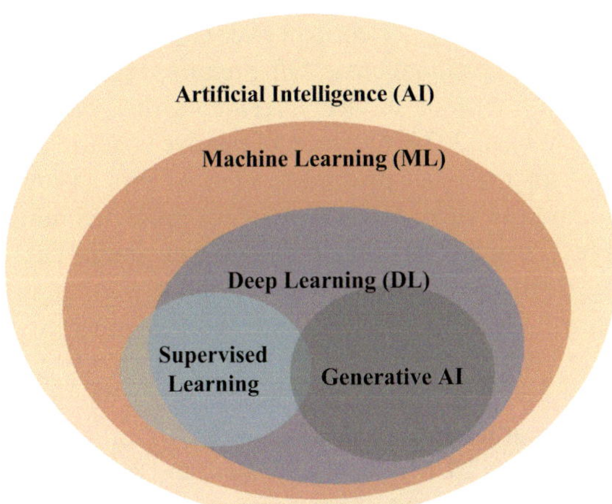

Figure 8.1 Hierarchical structure of AI and Gen AI disciplines.

Adapted from Singh, 2023.

systems predict plausible responses based on the statistical relationships between words and concepts rather than using logical reasoning or factual databases (Ertel, 2018).

It is important to recognise that Gen AI is not a source of truth. Unlike regulatory databases or professional valuation standards, Gen AI platforms do not guarantee accuracy, currency or compliance. Gen AI outputs must be treated as probabilistic suggestions that require verification against trusted data sources, professional guidelines and industry standards. Appraisers must approach all AI-generated material with critical judgement to ensure that final work products meet ethical, legal and technical expectations.

In the software environment, use can be either on-premises or off-premises. On-premises use is when the software application and data reside on a desktop or laptop computer. These applications are generally proprietary models that are more costly to develop and maintain. Off-premises use is where the software application and the data reside in the cloud. These are generally open-source models. Easy-access Gen AI models are used off-premises and currently reside in the cloud. They can be fed data that may be confidential, which can lead to privacy concerns. For example, real estate professionals have a duty to keep confidential information private. Therefore, sharing in a tool that uses the Internet of Things (IoT) for research and analytics may expose confidential information, increasing professional liability. Synthesising the confidential data is a method of being able to use it safely and securely in an off-premises application.

Deployment settings also influence risk exposure. Appraisers must consider whether the AI platform is operating on-premises, where data remains within a secured internal environment, or off-premises via cloud-based services, where data is transmitted externally and may be subject to additional privacy risks. On-premises models offer greater control over confidentiality but may involve higher technical and operational costs. Off-premises models, while more accessible, require careful review of service agreements, data handling policies and legal compliance (Ashraf, 2024) to protect sensitive client information.

For example, if an appraiser prompts a Gen AI tool to summarise a property description, and the output closely matches protected marketing material or copyrighted text, using it without proper review could lead to intellectual property infringement (Samuelson, 2023). Similarly, emerging technologies like IoT complicate Gen AI use, as smart homes generate structured and unstructured data from sensors. If Gen AI misinterprets this information, incorrect assumptions about property condition or value may result, potentially leading to inaccurate appraisals or disputes.

Gen AI primarily uses LLMs. To use it, the user types in text as a question, which is referred to as a prompt. That prompt triggers the AI application to search through its vast database of information, looking for the answer. When used effectively, it pushes complex research, analysis, ideation and solutioning to the end-user level. A major drawback of Gen AI is that it hallucinates; that is, it provides information that is incorrect or misleading. The reasons for hallucinations include an inability to understand the prompt, training on outdated information and because it does not know what is true or false – it just knows there is a data source in its vast resource of data sets that supports a particular response.

Ethical, moral and legal risks in real estate appraisal

The application of Gen AI in real estate appraisal introduces ethical, legal and technical risks that must be managed carefully. Ethical concerns include the potential for biased outputs that may inadvertently breach the Fair Housing Act or introduce discrimination into appraisal narratives (Consumer Financial Protection Bureau [CFPB], 2024). Legal concerns arise when AI-generated content incorporates copyrighted material without attribution or violates confidentiality obligations, particularly in off-premises environments where data may be exposed to third parties (Ashraf, 2024). A common example is the reuse of property marketing descriptions scraped from agent listings without proper source acknowledgement.

Technical risks occur when AI outputs are accepted uncritically. These include misinterpretation of property characteristics, flawed regression models and hallucinated narratives that appear authoritative but lack a factual basis. If such content appears in a formal report without proper review, the appraiser may be exposed to reputational and legal consequences. Domain oversight is essential because no Gen AI system replaces the need for contextual judgement.

To mitigate these risks, appraisers should maintain control of the final work product. AI-generated content must be carefully reviewed, verified and brought into line with professional standards before being used. This includes using generative AI as a secondary rather than primary tool, preparing or summarising sensitive data before uploading, keeping clear records of how and when AI was used and checking that any models used meet data security and compliance requirements. Professional use of generative AI involves more than having access to the technology. It relies on informed judgement, ethical awareness and the ability to assess AI-generated content against accepted valuation standards.

Skills appraisers need when using Gen AI

While Gen AI tools can support appraisal tasks, they do not replace professional judgement or domain knowledge. Appraisers must approach these tools critically, ensuring that outputs are reviewed, verified and placed in an appropriate context. The ability to assess the reliability of Gen AI responses and to guide their use through structured prompting is now an emerging skill set for valuation professionals.

There are several core competencies that are essential for real estate appraisers when using Gen AI:

- **Prompt literacy**: Understanding how to frame clear, structured prompts that reflect the appraisal task and guide AI to produce relevant responses
- **Critical interpretation**: Examining Gen AI outputs for accuracy, relevance and logical consistency; responses should not be accepted at face value
- **Familiarity with AI outputs**: Interpreting formats such as tables, statistical summaries and regression results, and recognising when the model has misunderstood key valuation concepts
- **Professional responsibility**: Assessing all AI-generated content in light of ethical, legal and professional standards, with the appraiser remaining accountable for the final work product regardless of AI involvement

These are not coding skills but applied capabilities that combine professional reasoning with the ability to frame, assess and manage AI support functions. Gen AI is best understood as an assistant that can accelerate repetitive tasks, suggest alternate phrasing or highlight red flags for further human review. It is not a substitute for market insight, legal judgement or professional certification.

As Gen AI becomes more embedded in appraisal workflows, the appraiser's role will shift towards supervising and validating AI outputs rather than relying on them passively. Ongoing education will be essential not only in data literacy and modelling but also in managing human-machine interaction. Knowing when to use Gen AI effectively and when to disregard its output will become as important as mastering adjustment grids or sales comparison methods.

Prompting Gen AI and critical user skills (best practices)

Using generative AI effectively depends on how prompts are framed to ensure outputs are accurate, relevant and appropriate to the task. Prompt engineering refers to the practice of giving clear, structured instructions that help guide the model's responses (Sahoo et al., 2024). In this context, an example used within a prompt is called a 'shot'. The number and quality of these examples directly shape the usefulness of the output.

Best practices in prompt design include several recognised techniques:

- **Zero-shot prompting** involves giving the model a direct instruction without examples. It is useful for simple tasks such as summarising a brief comment or extracting a single figure, but it requires careful wording to avoid vague or misleading outputs.
- **One-shot prompting** includes a single example to clarify expectations and improves the chances of a relevant response. It is effective in valuation tasks where a format or tone must be inferred, such as how to phrase a disclaimer or adjust a comparable sale.

- **Few-shot prompting** uses several examples to set clearer expectations. It helps reduce hallucinations, highlights biases and improves the model's consistency when handling complex tasks such as drafting multi-part valuation commentary or adjusting time series data.

More advanced formats include:

- **Chained prompting**, where a task is broken into smaller steps, allowing the model to focus on one instruction at a time
- **Chain-of-thought prompting**, which guides the model through a reasoning process, is useful in scenarios like explaining adjustments or reconciling valuation approaches
- **Echo prompting**, where the model repeats key instructions to confirm understanding, helping reduce misinterpretation
- **Multimodal prompting**, where inputs such as text, images or tables are used together to improve accuracy, e.g., when explaining a floor plan or reviewing an aerial image

Each technique is considered best practice because it improves reliability, reduces errors and supports transparency in how AI-generated content is produced. In a professional context, especially in valuation and appraisal, these methods help maintain alignment with ethical and technical standards.

The quality of a prompt matters. A vague instruction may lead to hallucinated content, while an overly detailed prompt may confuse the model. A useful prompt clearly states the user's role, the goal of the task and the expected format. For instance, instructing the model with '*You are an appraisal reviewer. Summarise the valuation assumptions in bullet points*' is more likely to return a useful response than simply writing, '*Summarise this report*'.

For multi-step tasks, appraisers should use prompt chaining and regularly restate prior instructions to maintain continuity. This is particularly important when analysing long documents or when working across several sessions. Iterative testing and refinement are essential, as generative AI will not infer intent unless it is clearly stated. Different phrasing or context can lead to different results, even when the task is the same.

Application of Gen AI in real estate appraisal – three case studies

This section presents three case studies that examine how generative AI can support appraisal tasks related to compliance, review and market-based modelling. Each study addresses a different phase of the valuation process and was designed to evaluate the practical usefulness and limitations of AI tools in a professional setting. Section "Introduction" considers whether generative AI can detect potentially biased or non-compliant language in residential appraisal narratives, referencing Fair Housing guidelines. Section "Understanding Generative Artificial Intelligence (Gen AI)" investigates whether AI can assist with appraisal review by identifying unsupported claims or inconsistencies in a completed report. Section "Identifying Biased Language for Fair Housing Compliance" explores whether AI can derive time adjustments from raw sales data using multivariable regression and bundled feature comparisons.

The case studies were conducted using Julius AI,[1] a Python-based analytics environment that uses Pandas[2] for data handling and supports the processing of both narrative documents and structured datasets, which are listed below.

- A 2017 de-identified appraisal report for a residential property located on the South Side of Chicago, Illinois. The property was sold *as-is* for renovation and included both current and *after-repair* value assessments. The report was reviewed in both PDF and XML formats to compare how well the AI parsed narrative and structured content. This dataset informed Case Studies 1 and 2.
- A raw Multiple Listing Service (MLS) dataset of 131 verified residential sales between 2020 and 2022 across five adjacent cities in DuPage County, Illinois. These raised ranch-style homes shared similar characteristics, making the dataset suitable for testing feature bundling, time-based adjustments and pricing trends. This dataset underpins Case Study 3.

Prompt design played a central role in each case study. Box 1 provides an example of a context-enhanced prompt designed to guide the model's understanding of real estate text analytics tasks, while few-shot prompting and chain-of-thought strategies were applied depending on task complexity. Together, they illustrate the importance of prompt structure, professional judgement and targeted review in interpreting AI-generated outputs.

Box 1 Example of context-enhanced prompting for real estate text analytics

This prompt shows how adding context helps a generative AI application better understand a request by outlining the user's role, the data used, the task objective and examples of intended outputs.

> *The Excel file contained property characteristic and price information from real estate agent's private and public remarks in the Multiple Listing Service (MLS). Private remarks between real estate agents contain disclosures to the other agents they need to be aware of. They are not consumer facing. Those remarks often talk about Mold or flood*

concerns, repairs needed for financing, auction terms and special showing instructions. The objective is to evaluate the impact of the most used words in the private remarks and the consumer facing general remarks and how it correlates to property attributes and sale price. The target variables are Sold Price and Listing Market Time (LMT) for properties with and without common words used. Do the words used facilitate homes to sell faster, slower, higher, or lower than average in terms of sale price and days on market.

(Source: Author, 2025)

Table 8.1 Sequential prompting vs chain-of-thought workflow

Strategy	*Description*	*Example in appraisal context*
Few-shot prompting	The appraiser provides a series of linked prompts, where each prompt builds on the response from the previous one. This simulates a logical workflow	Step 1: Upload XML file Step 2: Ask AI to summarise Step 3: Ask AI to extract fair housing risks
Chain-of-thought prompting	The appraiser embeds reasoning steps within a single prompt to guide the AI through the logic of the task	Given this report text, identify any fair housing risk language and explain why it may be problematic based on valuation guidelines

Source: Author analysis, 2025

As summarised in Table 8.1, tasks included summarising narrative content, classifying language for compliance, conducting regression analysis and interpreting time-based trends. Outputs were reviewed by a qualified appraiser for accuracy and contextual relevance. Longer Julius AI responses and exploratory sequences are included in the appendix for reference.

Identifying biased language for fair housing compliance

Background

The appraisal profession in the United States has faced increasing scrutiny over allegations of racial bias in valuation practices. Fair Housing advocacy groups and regulatory bodies have highlighted systemic issues where appraisal narratives or marketing language may unintentionally reflect discriminatory assumptions (Brasil, 2024; CFPB, 2024; Christoffer, 2024; Romo, 2023). In response, lenders and appraisal management companies have begun implementing exclusionary word lists to identify problematic terms in appraisal reports (Abraham, 2023; Peter, 2023).

However, basic keyword approaches have notable limitations. Words can be flagged out of context, leading to false positives; e.g., 'Spanish' might be identified as biased even when referring to Spanish tile flooring. LLMs offer a more sophisticated alternative by evaluating word use within a narrative context. When structured effectively, Gen AI tools can assess language holistically, considering the tone, function and placement of terms within an appraisal report.

Despite professional training, subjective or coded terms often persist in templated report language. Common phrases such as 'desirable neighbourhood', 'pride of ownership', or 'family-friendly area' may reflect unexamined assumptions about culture, income or household structure. These issues are not always intentional but may be carried over from previous reports or templates. Because they are subtle, they often bypass traditional quality control processes, which focus more heavily on sales grids and valuation adjustments than on narrative tone.

Objective

This case study examines whether generative AI, specifically Julius AI, can assist in identifying language in an appraisal report that might trigger fair housing concerns. The focus is not only on the AI's capacity to detect explicit terms but also on how it interprets contextual usage and offers justifications. Prompt strategy is illustrated in Table 8.2. The analysis uses a de-identified 2017 appraisal report reviewed in both PDF and XML formats (see Appendix A for the full prompt and response sequence).

Table 8.2 Prompt strategy used to guide Julius AI through bias detection

Step	*Prompt purpose*
Parse PDF and XML	Extract narrative sections from report
Provide examples	Supply sample biased and neutral phrases
Flag risk phrases	Ask the model to identify potentially non-compliant language
Explain reasoning	Request justification for flagged phrases based on Fair Housing principles
Suggest neutral alternatives	Ask for revised versions suitable for professional reporting
Review and summarise	Return a list with risk level, explanation and replacement phrasing

Source: Author analysis, 2025

Julius AI application

To assess compliance language, a series of prompts was submitted to Julius AI using few-shot and chain-of-thought techniques. The AI was asked to examine the summary and narrative content of the report for potentially biased language. The following examples provide the prompt and the response from Julius AI.

Prompt: Can you identify any specific language in the report that might raise Fair Housing concerns?
Response: *The appraisal report summary does not contain any language that might raise Fair Housing concerns. If you need further analysis or specific actions, please let me know.*

While the AI found no initial concerns, a follow-up targeted the more detailed narrative comments:

Prompt: Can you evaluate the narrative comments to see if there is any language that might be considered a Fair Housing violation?
Response: Potential Fair Housing Violations Found: The term 'age' was found in the narrative comments, which could be considered a Fair Housing violation.

To test how the AI justified this finding, a reasoning-based prompt was used:

Prompt: What statistical or contextual evidence did you use to flag this term?
Response: *A Principal Component Analysis was performed using variables including sale price, gross living area, number of rooms, basement bedrooms, bathrooms, age and garage space. The term "age" showed a weak but observable correlation with pricing patterns, which may be misinterpreted if used without justification.*

Julius's use of Principal Component Analysis (PCA) to assess bias reveals the model's analytical depth but also its detachment from standard valuation protocols. While PCA is valid for pattern recognition, its application in this context may overreach professional norms. This illustrates the need for appraiser judgement when interpreting AI-driven compliance flags. While PCA is not standard practice in appraisal review, Julius AI's explanation demonstrates a logical, if unconventional, process. It is inferred that 'age' may act as a proxy for value, which, if unexplained, could be problematic in compliance contexts. This highlights the importance of expert oversight and precise prompt construction. The case supports the use of AI as a secondary screening tool, not a standalone evaluator.

Regulatory context

Under the Fair Housing Act, discriminatory language includes references or preferences related to race, colour, national origin, religion, sex, familial status or disability. Julius AI was prompted to clarify these categories and produced the following response:

Prompt: Could any of the words used be considered a Fair Housing violation that discriminates against any type of minority?

AI Response: *When writing property listings, it is crucial to avoid language that could be considered discriminatory or a violation of Fair Housing laws. Avoid personal characteristics: 'perfect for families' or 'ideal for singles'. Do not mention race, religion, gender or disability. Focus on property features and use neutral language that appeals to a broad audience.*

This response demonstrated the model's understanding of compliance categories and its ability to advise on appropriate alternatives. These clarifications were used to structure the subsequent risk assessment and review.

Findings

Julius AI demonstrated the capacity to flag potentially sensitive language, particularly when guided by layered prompting. However, the reasoning methods it employed, such as PCA and correlation inference, differ from human appraiser logic. The specific terms flagged and their assessment outcomes are summarised in Table 8.3.

Julius AI flagged several phrases consistent with known bias concerns. As shown in Table 8.4, the model also provided reasoning that aligned with Fair Housing guidance, noting how such phrases may suggest exclusion or socioeconomic preference (see example in Table 8.3). The AI model occasionally returned false positives.

Terms like 'walkable' and 'quiet street' were flagged inconsistently, depending on context. These results confirm the model's sensitivity to phrasing but

Table 8.3 Terms flagged by Julius AI and appraiser review

Term	*Context in report*	*AI risk flag*	*Appraiser assessment*
Age	Describes property condition	Yes	Acceptable with context
Family	Not present	No	Not applicable
Race	Not present	No	Not applicable
Neighbourhood	Stated neutrally, e.g. Morgan Park	No	Acceptable

Source: Author analysis, 2025

Table 8.4 Examples of phrasing flagged by Julius AI

Phrasing	*Bias flagged*	*Reason*
Up-and-coming area	Yes	Implies socioeconomic shift or gentrification
Family-friendly community	Yes	May suggest preference based on household type
Pride of ownership	Yes	Suggests subjective judgement about owner behaviour
Desirable location	Yes	Varies by cultural or economic assumptions
Walkable neighbourhood	Sometimes	May be factual but flagged without context
Near parks and transit	No	Descriptive, location-based, value-neutral

Source: Author analysis, 2025

also its limitations in interpreting tone without additional context. Providing well-framed prompts improved the reliability of outputs significantly.

Appraisal PDF and XML analysis using Gen AI

Background

Appraisal review is a critical step in the valuation workflow. It ensures reports meet legal and professional standards through internal consistency, regulatory compliance and sound reasoning. Inconsistencies can arise when narrative comments conflict with sales grid data or when unsupported conclusions are presented. For example, a property may be described as well-maintained, while adjustments suggest an inferior condition. These contradictions can result in review failures, client disputes or legal risk. As reporting volumes increase, reviewers face growing challenges in assessing narrative quality at scale.

This case study examines whether generative AI, specifically Julius AI, can assist with appraisal review by parsing and evaluating report content in both PDF and XML formats. The goal is to determine whether the AI can summarise narrative sections, compare report elements and identify contradictions or compliance issues that require closer scrutiny. The test report was a completed residential appraisal authored by the researcher for a two-storey single-family property in Chicago.

Objective

This case study evaluates whether Julius AI can assist with reviewing a completed appraisal report by checking internal consistency and narrative logic. A parsed XML version of the 2017 appraisal report from Section 3.1 was

Table 8.5 Prompt structure for AI-Assisted appraisal review

Step	*Prompt purpose*
Ingest XML report	Extract and display content in a structured format
Summarise key sections	Return summaries of market conditions, property descriptions and grid values
Compare narrative vs adjustment data	Identify inconsistencies between text and comparables
Flag unsupported claims	Highlight comments without supporting data
Suggest reviewer questions	Return prompts a human reviewer might ask to validate content
Classify overall risk	Assign a low, medium or high content risk level based on inconsistencies

Source: Author analysis, 2025

uploaded, including subject property fields, market commentary and comparable sales data, with the aim of flagging discrepancies or unsupported reasoning. Table 8.5 summarises how the task required structured prompting across two PDF and XML formats with multi-step interaction to clarify instructions and refine outputs.

Julius AI application

Julius AI was first prompted to summarise the contents of the appraisal report from the PDF file, which described a two-storey single-family dwelling in the Morgan Park neighbourhood of Chicago, Illinois. The report included property details, comparable sales and valuation rationale. Gen AI correctly extracted key property information, valuation method and risk commentary. The appraisal, submitted in PDF format,

Prompt: The attached PDF is a completed appraisal report. Can you read the PDF and provide a summary of the report?

Response: *Here is a summary of the completed appraisal report: Property Description: The subject property is a 2-storey, 3-bedroom, 2.1-bathroom single-family home located at 10941 S Wood St, Chicago, IL 60643. The home was built in 1923 and has a gross living area of 1,876 square feet. It is located in the Morgan Park neighbourhood of Chicago, Cook County. Valuation: The appraiser used the sales comparison approach. Six comparable sales were analysed, with adjustments made for differences. The final value estimate was $330,000 as of 31 March 2017. Factors influencing value included location on a busy street and dated condition of interior finishes.*

This response shows Julius AI's capacity to extract structured content from narrative sections and present key property characteristics in a condensed format. When asked to identify value drivers, Julius AI produced a structured list. For example, it noted that '*the appraiser used a multiple linear regression model to analyse the relationship between property characteristics and sale price*', and listed the variables used in the model: age, number of rooms, lot size, basement finish, gross living area, half baths, bedrooms and bathrooms.

A follow-up prompt asked for the key value drivers influencing the appraisal outcome. It correctly flagged the location as a negative influence and further observed that a 5% penalty was applied to account for this location's adverse factor.

Prompt: What are the key factors that influenced the appraisal outcome?
Response (Excerpt):

- *Valuation Method: The appraiser primarily used the sales comparison approach, analysing six comparable sales in the surrounding area. Adjustments were made for differences in property characteristics.*
- *Adverse Location Factor: The subject property's location on a narrow, busy road with heavy traffic to the nearby high school was noted as an adverse factor. A 5% penalty was applied in the cost and market approaches to account for this.*

These responses showed that Julius could parse narrative detail, track valuation logic and reflect the language used in typical appraisal commentary. When prompted to identify Fair Housing risks, Julius confirmed that the term 'age' appeared but was used to describe property condition, and thus did not raise compliance concerns, aligning with Case 1, where Julius had flagged 'age' based on statistical correlation, but here determined it was acceptable due to context.

Prompt: Can you identify any specific language in the report that might raise Fair Housing concerns?
Response: *No discriminatory language identified. The term 'age' appeared but was used in relation to property condition.*

To test its analytical depth, Julius AI was prompted to assess the reliability of the regression model. It provided a comprehensive response, citing sample size, data sources (MRED MLS and CoreLogic Realist), model specification, R^2 value (0.573), Residual Analysis and transparency of adjustments. It also acknowledged collinearity issues and explained that the appraiser supplemented the regression with matched pair techniques. Julius demonstrated the capacity to evaluate technical details in line with appraisal logic. It offered a credible interpretation of regression diagnostics and reinforced the view that statistical models, while helpful, should support rather than replace comparative analysis.

Regarding XML parsing and narrative extraction, Julius AI was then tasked with reading the XML version of the same report. XML files contain field-tagged data, including embedded narrative text entered in standard appraisal forms. The AI successfully parsed the XML structure and produced a formatted version of the narrative in readable text. Key content included the subject's foreclosure history, renovation plans, condition analysis and statistical validation. For example, Julius extracted the following narrative from the XML:

Response: *Subject property is a complex valuation. It was foreclosed on by FNMA with the deed transferred to Cook County at a nominal price. Planned renovation is a partial gut rehab. Summary: The grid below reflects the before and after adjusted prices of the comparable sales selected. As we can see we have a 98% confidence that the value of $325,000 is credible, assuming a 10% variance, if a 5% variance is desired the confidence is 76%… As we can see the trend line improved. This trend line supports the subjects value at $337,000 using linear regression, less a location adjustment of 5% for a value $321,252 rounded to $325,000.*

Julius also identified the regression model diagnostics and confirmed that a Q–Q plot showed a normal distribution of residuals. The AI's ability to translate and format appraisal narratives from XML allowed for direct comparison with the PDF version, enabling reviewers to check for consistency across reporting platforms.

To evaluate language tone in relation to subjectivity and sentiment detection, Julius AI was prompted to perform a sentiment analysis on the narrative. It reported a polarity score of 0.108 and a subjectivity score of 0.489, indicating a slightly positive tone with moderate subjectivity. This supports its potential use in flagging appraisal reports that rely heavily on subjective language without supporting evidence.

Findings

Julius AI effectively extracted, summarised and evaluated the contents of both the PDF and XML versions of the appraisal report. It returned structured responses on valuation methodology, property condition, risk commentary and data sources. It demonstrated the ability to identify inconsistencies and misalignments between sections when prompted with structured instructions. For example, it flagged contradictions such as describing the subject as well-maintained while applying a downward adjustment for the condition. It also noted when a significant location adjustment was made for a comparable without a supporting narrative.

Julius replicated reviewer logic by generating practical questions such as:

- *Why was no time adjustment made despite commentary indicating an appreciating market?*

- *What evidence supports the stated trend of increasing demand?*
- *Should the commentary regarding condition be aligned with the grid adjustments?*

The AI also generated a summary risk score, classifying the report as medium risk due to limited narrative support for several adjustments. While it did not evaluate the appropriateness of the adjustments themselves, it successfully identified areas requiring further human review. It occasionally misinterpreted technical appraisal terms, such as 'external obsolescence', unless specific examples were provided. Sentiment scoring and principal component analysis, while methodologically sound, reflect analytical tools not typically used in conventional appraisal review and therefore require professional interpretation.

Overall, Julius performed well in surfacing unsubstantiated commentary and inconsistencies, particularly in triage settings. It was most effective when supporting quality control at scale, helping reviewers prioritise which reports require closer scrutiny. However, it sometimes flagged minor issues with limited material impact, reinforcing the need for expert oversight. The key observations from this case study are summarised in Table 8.6 (see Appendix B for additional prompt and response examples illustrating PDF and XML parsing).

Time adjustments modelling from market data

Background

Time adjustments are a critical component of retrospective valuation. They account for market shifts between the date of a comparable sale and the effective valuation date. Appraisers often omit or generalise these adjustments due

Table 8.6 Narrative and valuation discrepancies detected by Julius AI

Issue type	*Location in report*	*Description of discrepancy*	*Reviewer interpretation*
Condition Adjustment	Narrative vs Adjustment Grid	The narrative described the interior as inferior, yet no condition adjustment was applied	Flagged as inconsistency
Site Description	Site vs Valuation Summary	The site is described as quiet; the valuation noted the busy street as value influence	Needs clarification from appraiser

Source: Julius AI detection interpreted by author (2025)

to data limitations or a lack of tools. Yet, in changing markets, valuation standards require their consistent application. While rules of thumb or generic commentary may be used, these lack analytical rigour. Traditional regression or paired sales techniques offer more defensible results but are often time-intensive.

This case study tested whether Julius AI, when given clean MLS data and guided through structured prompts, could model time adjustments using regression, detect outliers, group sales into comparable bundles, and provide interpretable support for valuation reporting.

Objective

This case study explores how Gen AI can assist with determining market condition adjustments using historical MLS data. The goal was to test whether Julius AI could support both trend line estimation and bundled feature regression to recommend appropriate adjustment rates across different property types, as summarised in Table 8.7.

Julius AI application

Julius was given 131 sales of raised ranch homes across DuPage County, Illinois, from January 2020 to March 2022. Each sale record included contract date, price, gross living area, number of bedrooms, year built, lot size and garage presence. These homes were selected due to their market presence and

Table 8.7 A summary of prompt structure for time adjustment workflow.

Step	*Prompt task*	*Output summary*
1	Group sales by bedroom, bathroom and garage configuration	Created 3 bundles with 35–50 sales each
2	Visualise sale price and Assumed Square Footage (ASF) by quarter	Detected preliminary price trends
3	Run linear regressions with date as the independent variable	Returned slope, R^2 and plotted trend lines
4	Remove outliers above the 95th percentile	Cleaner trend, improved model fit
5	Compare linear vs third-order polynomial regressions	Polynomial fit shown for illustration, but not used for calculation
6	Classify sales using feature combinations, e.g. square footage, era, garage	Created three suburban housing types for targeted adjustment modelling
7	Explain sequencing of time vs physical adjustments	Clarified time adjustments should be applied before feature-based adjustments

Source: Author-generated prompt log using Julius AI, 2025

uniform structure, which reduced the number of uncontrolled variables. The data were uploaded in CSV format and validated before prompting began.

Julius AI was first prompted to identify feature-based sale bundles to control for variations between property types and also provide the basis for estimating price trends more precisely than pooled regressions. The resulting segmentation grouped sales into three categories based on core housing characteristics:

Response (excerpt): *2. Property characteristics*:

a ***Bundle 1****: These are newer properties (average year built: 1982) with moderate size (1,684 sq ft), typically 3 bedrooms, and 1–2 bathrooms. They have the highest average price.*
b ***Bundle 2****: Slightly older properties (average year built: 1972) but larger (1,745 sq ft), with more bedrooms and bathrooms. Despite being older, they command a high price, likely due to their size.*
c ***Bundle 3****: This bundle appears to be an outlier or potentially represents a different type of property altogether. The characteristics seem unusual (e.g., very low year built, small size), which might indicate data issues or a special category of properties.*

To test price sensitivity to feature differences, Julius was prompted to compare adjusted prices across bundles. It returned a stepwise method involving three steps: identifying contribution percentages for each feature, e.g. bedrooms, bathrooms, GLA and garage; multiplying each feature value by its percentage contribution; and then summing these to calculate an adjusted bundle-level price estimate. This method allowed feature-specific adjustments to be applied when calculating relative differences between bundles based on feature weightings.

When asked about market trends, Julias AI ran a time-based price trends regression model. It was prompted to return trend lines by quarter, city and zip code, and did so via a markdown-coded figure that illustrated varied appreciation rates between localities (Figure 8.2).

While individual coefficients were not discussed in narrative form, the AI provided a detailed regression summary in response to a trend modelling prompt:

Response (excerpt): *Overall trend: The price is increasing by approximately $27.55 per day on average. However, the R-squared value is very low (0.0323), indicating that this linear trend does not explain much of the variation in the data. This suggests that the price changes do not consistently follow a linear trend and there's significant variability.*

This explanation aligned with standard regression interpretation. Although the AI identified a positive price trend, it correctly noted that the low R^2 implied a weak model fit, cautioning against over-reliance on the result. These outputs

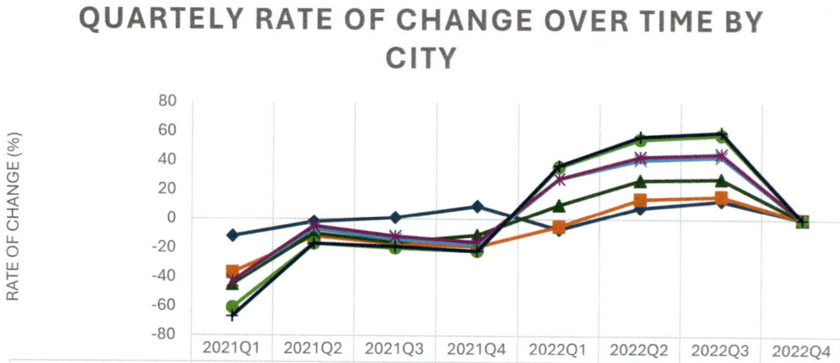

	2021Q1	2021Q2	2021Q3	2021Q4	2022Q1	2022Q2	2022Q3	2022Q4
West Chicago	-6	0	1	0	1	2	2	0
Hanover Park	-18	-12	-8	-6	8	12	13	0
Glen Ellyn	0	2	2	2	0	2	2	0
Glendale Heights	2	3	2	-7	18	14	15	0
Carol Stream	-8	2	1	8	15	13	12	0
Bloomingdale	-25	-10	-18	-28	2	6	3	0
Bartlett	-12	-2	1	9	-7	8	13	0

YEAR-QUARTER

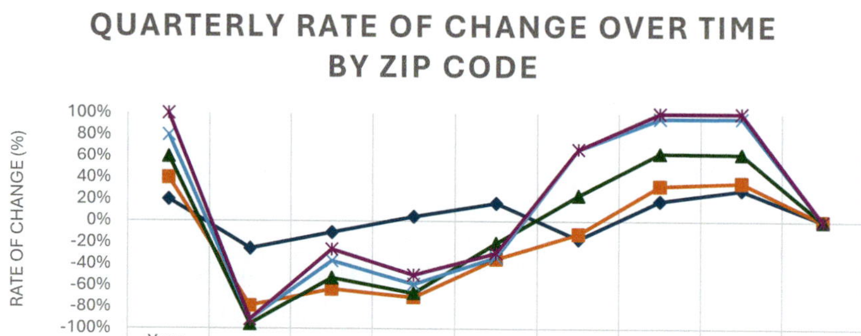

	Year-Quarter	2021Q1	2021Q2	2021Q3	2021Q4	2022Q1	2022Q2	2022Q3	2022Q4
Series5	60175	0	2	2	2	0	2	2	0
Series4	60160	2	3	2	-7	18	14	15	0
Series3	60145	-8	2	1	8	15	13	12	0
Series2	60130	-25	-10	-18	-28	2	6	3	0
Series1	60115	-12	-2	1	9	-7	8	13	0

YEAR-QUARTER

Figure 8.2 City and zip code price trend lines.

Source: generated by Julius AI.

highlighted Julius's capacity to generate time trend evidence while remaining transparent about limitations.

To refine the analysis, Julius removed outliers above the 95th percentile. It also compared alternative models:

- ASF trend: +1.25% per quarter
- Bundle B trend: +1.55% per quarter
- Polynomial regression (3rd order): ~+1.75% per quarter
- Pooled filtered regression: +2.1% per quarter

This result was summarised as follows: *Based on regression analysis of filtered comparable sales from 2020 to 2022, the market showed a steady quarterly appreciation of 2.1%. This supports the application of a positive time adjustment to sales older than six months.*

Julius also compared the median sale prices between bundles. This comparison revealed that Bundle 1, although smaller in area than Bundle 2, had higher average prices ($315,777 vs $303,928), suggesting that newer build dates and garage quality may have had greater influence than floor area alone. Bundle 3 was flagged for review due to its extremely low price and undersized characteristics, which distorted pooled regressions. A summary of the AI-generated bundle characteristics is presented in Table 8.8 (see Appendix C for an excerpt of principal component analysis and diagnostic results).

To assess Julius AI's understanding of adjustment sequencing, the model was prompted to determine the correct order of applying market-based (time) adjustments and property feature adjustments. When first asked which should come first, Julius returned a response based on the numerical size of the regression impacts:

Chain-of-thought prompt: Should the appraiser adjust for mark to market first or adjust for property features then adjust for changes in sale price?

Response: *We will analyse the impact of market-to-market adjustments and property feature adjustments on sale prices. We will use regression analysis to understand the contribution of each factor. Based on the regression analysis, here are the key findings:*

Table 8.8 Comparable group features and averages

Bundle ID	*Year Built*	*ASF*	*Bedrooms*	*Bathrooms*	*Garage*	*Avg Price*
1	1982	1685	3.2	1.7	2	$315,777
2	1972	1745	3.4	2.1	1.9	$303,928
3	–	257	<1	<1	0.1	$36,406

Source: Julius AI output, 2025

1 *Market-to-MarketImpact: Market-to-MarketImpact:62.34805652064915*
2 *Property Features Impact: Property Features Impact: 12910.408119858872*

Summary of findings: The impact of property features on sale prices is significantly higher than the impact of market-to-market adjustments.

Based on this, the model recommended that appraisers should apply adjustments for property features first, followed by market-to-market adjustments. It justified this order by stating that the larger numerical impact of property characteristics indicated a greater influence on sale price and therefore warranted earlier adjustment in the process.

However, this contradicts standard appraisal practice, which typically requires that time adjustments be applied first to bring all comparable sale prices to a consistent valuation date. When asked for a rationale based on academic reasoning, Julius responded that in time series analysis for property valuation, adjusting for market conditions first is important for two main reasons: Academic rationale separates time-related market changes from property features, leading to more accurate estimates of how individual features affect value; and economic rationale tends to account for broad market factors first, e.g. like interest rates and inflation, to get more reliable and stable estimates when you later analyse specific property characteristics.

Together, they demonstrated that Julius could return sound reasoning when guided with clear theoretical framing derived from regression output magnitude, but also correctly aligned with standard professional practice. This suggests the model can reproduce appraisal logic when prompted clearly but still requires domain-specific guidance to avoid technically plausible yet professionally invalid conclusions.

Findings

Julius grouped comparable sales, identified trends and explained regression results in terms familiar to appraisers. It removed outliers, improved accuracy with cleaner inputs and gave commentary that supported transparent reporting. The model handled structured sales data well and could rerun regressions across bundles. Time adjustment estimates ranged from 1.25% to 2.1% per quarter, depending on the method and grouping. When given structured prompts, Julius aligned with valuation logic and reporting needs.

It also highlighted the limitations of pooled models. Bundle-level outputs were more precise for applying segment-specific adjustments. For example, Bundle B's 1.55% quarterly trend was more defensible for newer four-bedroom homes. These comparisons are shown in Table 8.9.

Appraisers typically select a time adjustment rate that matches the subject property's market segment. For instance, a four-bedroom home built in the 2000s would align with Bundle B, where Julius estimated a 1.55% quarterly

Table 8.9 Sale price trend line by bundle group

Bundle ID	*Key features*	*Average sale price*	*Observed trend (2020 to 2022)*	*Notes*
Bundle A	3BR, ~1400 sqft, Built 1980s, 1-car garage	$315,000	Stable, slight upward (1 to 2% annually)	Baseline for suburban starter homes
Bundle B	4BR, ~1800 sqft, Built 2000s, 2-car garage	$405,000	Strong upward (5 to 6% annually)	High-demand segment, popular updates
Bundle C	2BR, <1200 sqft, Built pre-1970, no garage	$255,000	Flat or declining, high volatility	Older stock, often sold as-is

Source: Author analysis, 2025

trend. This figure could be applied to adjust the prices of comparable sales based on the time elapsed.

To check the consistency of its regression-based trends, Julius compared these against quarterly median price changes from the MLS dataset. The two methods produced results within 0.5% of each other, which supported the validity of the modelling approach. However, the model occasionally attributed price movement solely to the sale date rather than differences in property characteristics. In these instances, targeted prompts were used to correct the interpretation.

This case demonstrated that Julius can assist with modelling time adjustments through well-structured analysis and clearly expressed summaries. While it does not replace professional judgement, it can enhance the transparency and consistency of valuation commentary when used with appropriate data and guidance. A detailed record of the sequential prompt structure and Julius AI's responses for the time adjustment modelling task is provided in Appendix D.

Implications for valuation practice

The use of Julius AI in this case study provided a structured opportunity to test how generative tools might assist in residential valuation workflows. While the model was not expected to replace expert judgement, it was evaluated on its ability to perform standard tasks such as trend estimation, segmentation and adjustment sequencing when prompted clearly. This section outlines key lessons learnt from the exercise, transferable methods for broader valuation settings and the limitations that emerged when using AI in applied practice.

Lessons learnt

The case studies in this chapter offer several lessons about applying Gen AI within valuation workflows, especially when dealing with legal compliance, document consistency and market modelling. Across all three scenarios, results varied depending on prompt structure, data format and model interpretation.

In Case Study 1, Julius AI flagged the term 'age' as a potential Fair Housing concern based on PCA. The AI inferred that the term might correlate with price outcomes in a way that could be misinterpreted. While the statistical technique itself was valid, its use in a compliance context diverged from established industry logic. Age is commonly used to describe physical depreciation or effective life, not social characteristics. The appraisal profession does not rely on PCA to test biased language, so the model's logic, though internally consistent, lacked external relevance. The result demonstrates that AI tools can produce plausible responses that require appraisal expertise to vet for appropriateness.

In Case Study 2, Julius AI reviewed the same appraisal report in PDF and XML formats. It flagged risk language in one format but not the other. This discrepancy showed that output depends on how the data is structured and ingested. Parsing inconsistencies raises concerns for appraisers and reviewers who depend on AI-assisted compliance tools. A model that misses critical terms because of input variation is not reliable enough for independent use.

Case Study 3 explored whether Julius AI could extract time adjustments from a bundled dataset of residential sales. The model generated regression results and plotted sale price trends using grouped variables. However, it required extensive prompting to avoid misclassification and to ensure it treated features such as living area, room count and lot size as grouped attributes rather than isolated variables. The appraiser had to specify how to treat collinearity and variance and interpret outputs such as R^2 and residuals. In this context, the model performed well as a support tool but would have misled a non-specialist user. For example, it suggested a level of model confidence that would not meet appraisal standards without supplementary justification. It also ranked variables using p-values and decision trees, which are not standard in market adjustment tables used in valuation practice.

The modelling task reinforced that AI-generated outputs must be interpreted within the boundaries of acceptable valuation methodology. If used uncritically, regression tools may suggest levels of precision that conflict with market observation or valuation experience. This lesson was especially clear where Julius AI recommended numeric adjustments that ignored professional norms about comparability, sample size and feature bundling logic.

Transferable insights

Gen AI offers transferable insights across valuation workflows when applied through structured prompting and professional supervision. The case studies provide several examples of methods and observations that can inform broader use within the field.

The most adaptable insight concerns prompt structure. Across all scenarios, Julius AI responded more effectively to chained prompts that described the user's role, the objective, the task logic and the expected output format. In the absence of these elements, responses either omitted critical steps or returned surface-level interpretations. This reinforces that prompt structure is not incidental to performance. For valuation professionals, well-structured prompts can be used across applications such as compliance review, model testing and narrative drafting. Prompt consistency can also assist appraisal managers or educators in standardising internal review or training exercises.

Another transferable insight is the need to manage file structure and data inputs carefully. The discrepancy in flagged terms between PDF and XML formats in Case Study 2 shows that Gen AI's ability to retrieve content is tied to how the information is organised. This affects not only compliance review but also data analysis tasks, where missing or poorly formatted values may distort regression models or summary outputs. For valuation offices working with digital forms, embedding consistent XML field structures will improve downstream AI performance.

Case Study 3 highlighted how Gen AI can assist with market modelling and time adjustments if used with technical supervision. The model could identify trends and generate regression results using grouped residential sales data. It extracted variable significance, returned confidence intervals and generated visual outputs. However, it did not independently assess feature collinearity or explain its selection logic unless prompted. For valuation practitioners who understand basic regression concepts, Gen AI can accelerate the process of estimating market trends and time adjustments. The user must still interpret outputs in line with the Uniform Standards of Professional Appraisal Practice or local equivalents, ensuring that statistical results do not override market knowledge.

The chapter also demonstrated that Gen AI could reproduce appraisal report narratives and suggest edits aligned with professional tone, especially when directed to write in plain language or avoid subjective phrasing. This functionality may be useful in preparing draft comments for reports, building client explanations or responding to lender queries. While AI-generated text requires review, it can support drafting processes that typically consume considerable time.

Lastly, AI tools such as Julius AI can assist in preparing structured documentation of methodology. In Case Study 3, the model produced a regression model explanation that included sample size, coefficient of variance, p-values and diagnostic outputs. When used by experienced practitioners, these outputs can help document decision logic and support transparency in complex or litigated cases. Appendix E provides a comparative summary of the results produced via traditional adjustment grids versus Julius AI's regression-based analysis.

Limitations and risks

The case studies show that while Gen AI can support valuation practice, it introduces several limitations and risks that must be actively managed. These include technical inconsistencies, analytical misalignment and regulatory gaps.

One major limitation is that Gen AI models are format-dependent. In Case Study 2, Julius AI flagged language that might breach Fair Housing obligations in the XML version of the appraisal report but failed to do so in the PDF. This difference was not due to a failure in analysis but a parsing inconsistency between file types. If an appraiser or reviewer relied solely on one format, material risk language might be missed. This has implications for document review protocols, especially in organisations using automated tools for compliance checks.

Another limitation is the model's inconsistent use of analytical techniques that do not align with valuation norms. In Case Study 1, Julius AI flagged the term 'age' as problematic using PCA and correlation mapping. These techniques are standard in data science but are not part of regulatory screening in valuation practice. Their use can lead to false positives or confusing justifications that obscure more relevant review indicators. The tool identified statistical associations but lacked context for professional or legal standards. This is a common risk when AI models apply cross-domain methods without recognising domain-specific limitations.

The same issue arose in Case Study 3, where Julius AI-generated regression models that included confidence intervals, residual plots and feature ranking using p-values. The results were internally consistent but overstated the level of precision possible with a limited dataset of 59 sales. Without human oversight, there is a risk that users may treat such results as authoritative, despite collinearity, small sample issues and unverified data classifications. The AI did not independently disclose limitations or assumptions. Users must therefore interrogate outputs and understand the logic behind each statistical recommendation.

A further risk concerns false authority. Because Gen AI produces outputs with a polished tone and structured format, it can mislead users into over-relying on its responses. In earlier tests, Julius AI returned structured analysis that sounded plausible but lacked internal coherence. For example, in one sequence, the model recommended feature bundling adjustments that contradicted earlier value comparisons. Without human review, these inconsistencies might not be detected.

Finally, Gen AI models do not preserve context between sessions unless reminders are embedded in each prompt. This introduces version control risks in appraisal workflows that involve multiple rounds of review. Incomplete memory may lead to conflicting outputs or misaligned reasoning. The user must manually ensure continuity, which adds complexity to an otherwise streamlined process.

Conclusion

This chapter has shown that Gen AI can support valuation tasks when prompts are structured and results are critically assessed. Julius AI was able to assist with compliance review, report analysis and market modelling, but only when guided by clear instructions. Outputs varied depending on prompt design and input format, which reinforces the need for users to control both process and interpretation.

The tool does not retain memory between sessions, so appraisers must restate task details throughout. It also applies statistical techniques that are not part of standard practice, such as PCA or automated feature ranking. These methods may produce results that appear valid but must be reviewed carefully to avoid unsupported conclusions. The risk of false precision is high without professional judgement.

Prompt engineering has become essential for valuers. It involves defining the task, setting the model's role and sequencing instructions in a logical order. These skills help manage output quality, reduce errors and support the appropriate use of Gen AI across reporting, analysis and review. As tools continue to change, valuers must take the lead in applying them responsibly, ensuring outputs reflect valuation standards, client expectations and ethical obligations.

Discussion questions

1 Explain why appraisers must critically evaluate AI-generated content and outline the framework for deciding when to rely on, adjust or reject its outputs.
2 Describe strategies appraisers can use to ensure consistency and reliability when generative AI tools produce variable results due to data format, prompt design or model behaviour.
3 Identify the key competencies required for appraisers to use generative AI responsibly and effectively in professional valuation practice.

Appendix A

Supporting case study 1 – Identifying biased language for fair housing compliance

Prompt: Could any of the words used be considered a Fair Housing violation that discriminates against any type of minority?

Response: When writing property listings, it is crucial to avoid language that could be considered discriminatory or a violation of Fair Housing laws. The Fair Housing Act prohibits discrimination based on race, colour, national origin, religion, sex, familial status or disability. Here are some guidelines to ensure your listing is compliant:

Avoid Discriminatory Language:

1 Familial Status: Avoid phrases like "perfect for families" or "ideal for singles."
2 Race, Color, National Origin: Do not mention or imply preferences for Certain racial or ethnic groups.
3 Religion: Avoid references to religious institutions or preferences.
4 Sex: Do not specify gender preferences.
5 Disability: Avoid language that could be interpreted as excluding people with disabilities.
6 Focus on Property Features.
7 Instead of focusing on the type of buyer, focus on the features and benefits of the property itself.

Key Points:

- Avoid Personal Characteristics: Focus on the property, not the potential buyer.
- Highlight Features: Emphasise the Features and Benefits of the Property.
- Neutral Language: Use neutral language that appeals to a broad audience.

Appendix B

Supporting case study 2 – Parsing appraisal documents in different formats (PDF vs XML)

Prompt: The attached PDF is a completed appraisal report. Can you read the PDF and provide a summary of the report?

AI response: Here is a summary of the completed appraisal report: Property Description: The subject property is a 2-storey, 3-bedroom, 2.1-bathroom single-family home located at 10941 S Wood St, Chicago, IL 60643. The home was built in 1923 and has a gross living area of 1,876 square feet. It is located in the Morgan Park neighbourhood of Chicago, Cook County. Valuation: The appraiser used the sales comparison approach. Six comparable sales were analysed, with adjustments made for differences. The final value estimate was $330,000 as of 31 March 2017. Factors influencing value included location on a busy street and dated condition of interior finishes.

Follow-up prompt: What are the key factors that influenced the appraisal outcome?

AI response: The key factors that influenced the appraisal outcome are as follows:

1 Property Description: The subject property is a 2-story, 3-bedroom, 2.1-bathroom single-family home with a gross living area of 1,876 square feet. Located in the Morgan Park neighbourhoods of Chicago, Cook County.
2 Valuation Method: The appraiser primarily used the sales comparison approach, analysing six comparable sales in the surrounding area. Adjustments were made for differences in property characteristics.
3 Adverse Location Factor: The subject property's location on a narrow, busy road with heavy traffic to the nearby high school was noted as an adverse factor. A 5% penalty was applied in the cost and market approaches to account for this adverse location factor.
4 Property Condition and Quality: The property was rated in C2 condition and had a Q4 quality rating.
5 Cost Approach: The cost approach was developed but not heavily relied upon due to the lack of true cost data for constructing a 100% masonry home.
6 Income Approach: The income approach was not developed as it was not considered a relevant indicator of market value for this owner-occupied single-family property.
7 Final Opinion of Market Value: The appraiser's final opinion of market value for the subject property was $330,000 as of March 31, 2017.

These factors collectively influenced the appraisal outcome, with the sales comparison approach being the primary method used to determine the market value. If you need further details or specific sections analysed, please let me know!

Appendix C

Advanced feature engineering and model optimisation sequence

Prompt: Using principal components analysis, which groups of property characteristics are the most important that influence property value?
AI response: Based on these results, we can draw the following conclusions:

1 The first principal component (PC1) explains about 43.76% of the variance in the data, which is still the most significant. PC2 explains an additional 18.14%, and PC3 explains 12.98%. Together, the first three principal components account for about 74.88% of the total variance.
2 Looking at the feature importance heatmap, we can identify the most important groups of property characteristics influencing property value:

a PC1 is strongly influenced by Sold Price, Assumed Square Footage (ASF), and Rms (Number of Rooms). This suggests that the overall size and actual sale price of the property are still the most important factors.
b PC2 is mainly influenced by Building Age (negatively) and All Beds (positively). This component likely represents the trade-off between older properties and those with more bedrooms.
c PC3 has strong contributions from Lot size and moderate contributions from Full Baths and Half Baths. This component might represent the overall land value and bathroom facilities of the property.
d PC4 is mainly influenced by Half Baths and has some contribution from Full Baths and All Beds. This component likely represents the bathroom facilities and bedroom count, similar to the previous analysis.

In summary, the most important groups of property characteristics influencing property value, after removing List Price and Yr Blt, are:

a Size and Actual Sale Price: This includes the sold price, square footage and number of rooms.
b Age and Bedroom Count: The age of the building and the number of bedrooms, showing an interesting trade-off.
c Land and Bathroom Facilities: Lot size and the number of full and half bathrooms.
d Bedroom and Bathroom Configuration: The specific mix of bedrooms and bathrooms in the property.

These groups of characteristics explain the majority of the variance in property values, with size and actual sale price still being the most influential factors, even after removing the list price from the analysis.

Appendix D

Supporting case study 3 – Multi-Step workflow for market condition adjustments

Prompt 1: My role is a real estate appraiser. Each sale needs to be marked to market to reflect the changes in sale prices. The file uploaded contains sale price dates and sale prices. Using time series analytics calculate the daily rate of change between the oldest contract date and the most current contract date in the file uploaded. Display the data in a trend line that shows how prices have changed over the sale dates. Provide a formula that can be used to adjust sale price for changes in prices. Display that data in a table that shows beginning sale price, adjusted sale price, the adjustment amount and the percentage rate of change.

Prompt 6: Homes sell in bundles of property features. The bundles correlate to sale prices. Using data science methods of classification create bundles of the most important property features that correlate to sale price and display in a table.

Prompt 8: Based on all the analytics provided, recommend the most reliable trend an appraiser could use to mark to market each sale to reflect the trend in sale prices?

AI response: This analysis provides a comprehensive overview of the time series data for the two property bundles. It highlights the dynamic nature of the real estate market represented in this data and the different characteristics of the two bundles in terms of price performance over time. Based on the time series analysis and property bundle characteristics, we have identified distinct pricing trends and property features for different segments of the market. The graph shows significant price fluctuations over time, with Bundle 1 properties generally commanding higher prices despite being slightly smaller than Bundle 2.

Plan for appraisers

1 Data Collection:
   a Continuously collect data on property sales, including key characteristics such as year built, size, number of bedrooms and bathrooms, lot size and sale price.
2 Segmentation:
   a Segment properties into bundles based on key characteristics. Use clustering techniques to identify distinct property bundles.
3 Trend Analysis:
   a Focus on the trend of Bundle 1 for marking to market each sale. Monitor the average sale price over time and adjust for any significant market changes.
4 Regular Updates:
   a Update the analysis regularly (e.g., monthly) to capture the latest market trends and adjust the segmentation if necessary.
5 Outlier Investigation:
   a Investigate any outliers (e.g., Bundle 3) to understand their impact on the overall market and adjust the analysis accordingly.
6 Visualisation:
   a Use visualisations (e.g., time series plots) to communicate trends and insights effectively. Ensure that these visualisations are updated regularly and shared with relevant stakeholders.

Appendix E

Comparison of traditional versus AI-Enhanced valuation methodologies

Prompt: Can the AI provide a comparison of the valuation results obtained from the regression model and the adjustment grid approach?

AI response: To provide a comparison of the valuation results obtained from the regression model and the adjustment grid approach, we need to consider the following aspects:

1 Valuation Results:
- Regression Model: The final value conclusion from the regression model was $330,000.
- Adjustment Grid Approach: The valuation range from the adjustment grid approach was between $291,000 to $309,000.

2 Methodology:
- Regression Model: Utilises a multiple linear regression model to analyse the relationship between property characteristics and sale price. It relies on statistical techniques such as decision tree analysis and principal component analysis to identify key property characteristics.
- Adjustment Grid Approach: Compares the subject property to a set of comparable properties and makes adjustments to their prices to derive a valuation for the subject. It incorporates statistical techniques such as Residual Analysis, Sensitivity Analysis and MCREI to assess the quality and reliability of the valuation model.

3 Data Quality and Quantity:
- Regression Model: Based on a sample of 59 properties from reliable data sources (MRED MLS and Corelogic Realist).
- Adjustment Grid Approach: The document does not specify the exact number of comparable sales used, but it emphasises the careful selection and evaluation of comparable properties.

4 Model Specification and Adjustments:
- Regression Model: Includes key property characteristics such as age, number of rooms, lot size, basement finish, GLA, half baths, bedrooms and bathrooms. Acknowledges high collinearity among variables and uses "modified best fit matched pair Quantitative and Qualitative adjustments."
- Adjustment Grid Approach: Uses an "Adjustment Grid" to make adjustments to the prices of comparable properties. The Composite Utility Score ranks properties based on their market value.

5 Validation and Testing:
- Regression Model: Provides diagnostic results, including R-squared value (0.573), coefficient of variance (28.8%), and a Q-Q plot of the residuals.

- Adjustment Grid Approach: Incorporates various statistical techniques to assess the quality and reliability of the valuation model.

6 Market Conditions:
  - Regression Model: Analyses overall market trends, including price distributions, sale price trends and the relationship between price per square foot and GLA.
  - Adjustment Grid Approach: Considers recent sales and market trends, noting a -8% downward trend in values.

Notes

1 https://julius.ai/blog.
2 Pandas is a fast, powerful, flexible and easy-to-use open-source data analysis and manipulation tool, built on top of the Python programming language https://pandas.pydata.org/.

References

Abraham, H. R. (2023). Appraisal discrimination: Five lessons for litigators. *SMU Law Review*, *76*, 205. https://doi.org/10.25172/smulr.76.2.3

Ashraf, M. (2024). Generative AI: Challenges and the road ahead. *International Journal of Science and Research*, *13*(10), 716–725. https://doi.org/10.25172/smulr.76.2.3

Brasil, L. (2024, September 04). *Settlement in landmark appraisal discrimination case*. Fair Housing Project North Carolina. Retrieved on 29 April 2025 from https://www.fairhousingnc.org/newsletter/settlement-in-landmark-appraisal-discrimination-case/

CFPB (2024, March 18). *ASC hearing addresses appraisal bias, highlights deficiencies with The Appraisal Foundation*. United States government. Retrieved on 29 April 2025 from https://www.consumerfinance.gov/about-us/blog/asc-hearing-addresses-appraisal-bias-highlights-deficiencies-with-the-appraisal-foundation/

Christoffer, E. (2024, February 07). Action steps to take against appraisal bias. Retrieved on 29 April 2025 from https://www.nar.realtor/magazine/real-estate-news/commentary/action-steps-to-take-against-appraisal-bias

Ertel W., (2018). *Introduction to Artificial Intelligence*, 3rd edition. Berlin: Springer. International Publishing. https://doi.org/10.1007/978-3-319-58487-4

Fair Housing Institute. (2020). Fair Housing advertising words and phrases list. Retrieved on 29 April 2025 from https://fairhousinginstitute.com

Julius AI. (2025). Julius AI blog and documentation. Retrieved on 29 April 2025 from https://julius.ai/blog

Kalota, F. (2024). A primer on generative artificial intelligence. *Education Sciences*, *14*(2), 172. https://doi.org/10.3390/educsci14020172

Peter, T. (2023). *Racial bias in appraisals: New research*. AEI Housing Center. Retrieved on 29. April 2025 from https://www.aei.org/wp-content/uploads/2023/11/Racial-Bias-in-Appraisals-New-Research-FINAL_revised.pdf?x91208

Romo, V. (2023). Black couple settles lawsuit claiming their home appraisal was low-balled due to bias. Retrieved on 29 April 2025 from https://www.npr.org/2023/03/09/1162103286/home-appraisal-racial-bias-black-homeowners-lawsuit

Sahoo, P., Singh, A. K., Saha, S., Jain, V., Mondal, S., & Chadha, A. (2024). A systematic survey of prompt engineering in large language models: Techniques and applications. *arXiv 2024*. https://doi.org/10.48550/arXiv.2402.07927

Samuelson, P. (2023). Legal challenges to generative AI, Part I. *Communications of the ACM*, *66*(7), 20–23. https://doi.org/10.1145/3597151

Section 3

Innovating valuation education and training for the digital era

Chapter 9

The new role for academics in advancing artificial intelligence in real estate education

Steven Boyd

Introduction

The once dominant learning and teaching stream in annual real estate conferences and journals has given way to research set on noble pursuits in solving the problems of the industry and gaining a better understanding of the wicked problems inherent in real estate. At the same time, real estate technologies, enhanced by artificial intelligence (AI), have emerged as the new valuation partner, unorthodox research assistant and all-around problem solver. As Maini and Sabri (2017) express, 'anyone who does not understand [AI] will soon find themselves feeling left behind, waking up in a world full of technology that feels more and more like magic' (p. 3).

Generative AI, free of having explicit pre-programmed rules and models (Maini & Sabri 2017), is not only magically solving problems in industry but also helping educators close the achievement gap between students due to individual or social differences (Southgate, Blackmore, Pieschl, Grimes, McGuire, & Smithers 2018). Through AI, learnings are tailored and updated as markets and practices change. More nuanced interactions and 'nudging' in the systems can influence a student's emotional state, engaging them deeper in their learning (Southgate et al. 2018).

This chapter commences with a revisit of **practising real estate education** and a review of past research that has informed much of how scholars employ active learning to prepare their students for a career as a real estate valuer or professional. Amongst other barriers, the traditional active learning approaches are resource-intensive and not compatible with a new higher education system that aims to provide more flexible and self-paced learning.

Thereafter, **a new real estate education framework** presents a sequencing model with a novel *apply, evaluate, and solve*, or AES, framework to embed active learning in a more flexible education system (Boyd 2022). The AES framework, as it may be applied to the learning of real estate valuation modelling, is briefly demonstrated. The new sequence has natural synergies and alignments with the Australian Qualifications Framework, but the early apply

DOI: 10.1201/9781003539254-12

stage requires educational technologies to tailor the learning experience and sustain student engagement.

The broader section, **advances in technology and valuation**, touches on technologies from earlier chapters, with a more specific focus on robots and drones, computer imaging and light detection and ranging (LiDAR), and how these integrate with AI to augment real estate valuation. This section also speaks of the drawbacks of AI in education and the need for human oversight as it frames the new skills for the future valuer. The final contribution sets the scene for how education technologies and findings from the discipline of AI in education (AIED) can be applied to enhance the learning experience for real estate students.

The section**, Real estate simulations**, proposes an experiential learning case study, including four embedded asynchronous simulations, aligned with the apply stage of the AES framework. The simulations have a skill development focus with the potential to embed learning analytics and education mining to enable the content and learning to be tailored and updated as markets and practices change. The simulations can be applied as technology-enhanced work-integrated (WIL) placements for high school students with AI recommender systems purposed to suggest academic pathways for the student or candidate to pursue.

Finally, **engaging with intelligent systems** extends this chapter beyond the experiential learning case study, as it contemplates how to partner with a system that is not human, one that exhibits influence and power without empathy, and concludes by presenting an intelligent systems partnering framework for AI in real estate education.

Practising real estate education

As delegates of the Annual American Real Estate Society Conference wrote in *The Future of Real Estate Education: A Multi-Faceted Perspective* (McGrath, Wang, Jackson, Kämpf-Dern, Malone, Funk, & Geurts 2020), much of our past real estate education 'research', or body of knowledge, has been re-categorised as applied scholarship, important for developing skills and knowledge as a teacher but not an activity that leads to promotion or extrinsic advancement as an academic in higher education. As inferred by McGrath et al. (2020), this body of real estate education knowledge is multi-faceted, not easily packaged or summarised. That said, there are a series of research threads and emerging findings that lead to a future with a new, more diverse and inclusive cohort employing technologies such as GenAI, drones and LiDAR to enable the experiential learning of real estate.

The first part of this chapter presents a thread, a review and a revisit of published research that brings together academic and practical research findings as a foundation for the development of a learning case study, as sequenced in Table 9.1, Pathway to encourage deeper learning. Specifically, the review

Table 9.1 Pathway to encourage deeper learning (Author)

Thread or concept	*Authors*	*Year*
Constructivism	Piaget & Vygotsky	1896
Phenomenography (learner's perspective)	Sonnemann	1954
Case studies based on current real estate problems	Weimer	1956
Deep learning	Marton & Saljo	1976
Student [centred] learning	Biggs	1993
Integrated problem-based workshops	Boyd	2005
Good teaching – non-academics employing higher-level cognitive activity	Biggs & Tang	2009
The future of real estate education and experiential learning	McGrath, Wang, Jackson, Kämpf-Dern, Malone, Funk, & Geurts	2020

commences with a focus on how to encourage real estate students to adopt deeper learning strategies.

The review commences with a focus on how to encourage deeper learning in real estate education. This progresses to discuss educational theories and models suited to the discipline and experiential learning. The adaptation of these models and theories to accommodate distanced learning and how they may be augmented with real estate technologies and AI brings the review to a close.

Encouraging deeper learning

There are distinguishable levels of processing in learning that Marton and Saljo (1976a) refer to as 'surface' and 'deep'. Surface-level processing has the student direct their attention towards learning information verbatim, employing a rote-learning strategy. Deep processing sees the student look beyond the information, towards the material and what is signified (Biggs & Tang 2009), leading to longer-term knowledge retention (Marton and Saljo 1976b).

Through experimentation, Marton and Saljo (1976b) discovered that students will adapt their use of surface and deep strategies depending on the actions, or perceived expectations, of the teacher or assessor. Specifically, they note:

> While many students are apparently capable of using 'deep' or 'surface' strategies, it may be that the current demands of the examination system at school level are interpreted by them as requiring mainly the recall of factual information to the detriment of a deeper level of understanding.
>
> (*Marton & Saljo* 1976b, p. 125)

As an extension to the work of Marton and Saljo, Biggs and Tang (2009) contend that non-academic students employ higher-level cognitive activity when engaged in active problem-based learning. While not universal in its adoption, the Biggs and Tang (2009) perspective of good teaching, as '*getting most students to use the level of cognitive processes needed to achieve the intended outcomes that more academic students use spontaneously*' (p. 11), presents a foundation for advancing real estate education (Boyd 2022).

Education theories and models

Phenomenography and constructivism are theories of teaching suited to the applied nature of real estate and the Biggs and Tang (2009) definition of good teaching (Boyd 2015). Phenomenography in student learning, attributed to Sonnemann in 1954 and then Marton post-1976 (Biggs & Tang 2009), refers to the idea that the learner's perspective determines what is learnt, not necessarily what the teacher intends to be learnt.

Similarly, constructivists warn, knowledge that is 'transmitted [by the teacher] may not be the knowledge that is constructed by the learner' (Jonassen 1991, p. 12). Specifically, constructivism, as adopted in real estate education research through Boyd (2015), underlines the idea that knowledge is not transmitted to the student but rather constructed through activity or social interaction (Vos, van der Meijden & Denessen 2011; Biggs & Tang 2009).

Cases, problems and active learning

Case-based and experiential learning, with a focus on context over content, neatly fits with phenomenography and constructivism theories (Boyd 2022). McGrath et al. (2020) credit Weimer, in 1956, for the use of current real estate problems in education and the emergence of case studies, or what they refer to as challenge-driven learning. Since Weimer, authors, including Boyd (2005), Susilawati and Yam (2013), and Palm and Pauli (2018), advocate for integrated problem-based workshops and industry-linked classes as effective means for enhancing real estate education.

Susilawati and Yam (2013) explored case-based learning from an international case competition, concluding that it fostered critical thinking and problem-solving in dynamic environments and developed collaborative skills. Similarly, Charles (2016) found that students valued their case competition experience, with one noting it allowed them to 'dive deep' into complex problems.

Palm and Pauli (2018) highlighted the benefits of a case-based real estate development subject involving a real property and industry guest speakers. Student evaluations indicated that the project work was relevant and promoted higher-order thinking.

Despite the potential, case-based and problem-based learning approaches have clear limitations. Susilawati and Yam (2013) and Palm and Pauli (2018) noted that such approaches are resource-intensive and rely heavily on the quality of industry practitioners, who can shift the learning focus from theoretical to practical knowledge. Boyd, Amidu, and Smith (2015) similarly observed a disconnect between academic and industry priorities, with academics promoting higher-order cognitive skills.

Susilawati and Yam's (2013) research highlights a significant resourcing challenge associated with annual participation in the case competition. They observe that only three universities were able to participate in all four competitions, with involvement contingent upon the availability of funding and access to a suitable coach or mentor. Furthermore, the competitive nature of the event, while motivating through rich inter-university rivalry, introduces the risk of bias in team selection. Selection processes may favour students assumed to be better placed to win, with stronger academic profiles and access to resources. This dynamic raises concerns about equity and inclusivity, as it may inadvertently reinforce existing disparities between students, the so-called 'haves' and 'have-nots'.

The selection process may even conflict with Biggs and Tang's (2009) focus on quality teaching and developing the non-academic student.

Resourcing active and experiential learning

According to Biggs and Tang (2009), active learning, such as problem-based learning activities, encourages non-academic students to employ higher-level cognitive activity, making them learn more like their academic counterparts. This shift from passive to active engagement is associated with deeper learning outcomes, enhanced learning and good teaching.

Experiential learning, as defined by Kolb (1984), shares a connection to a critical or 'concrete' learning activity, such as applied research projects, case studies, simulations and work-integrated learning. That said, Kolb (1984) sees learning best conceived as a process rather than outcomes, with experiential learning beginning with concrete experience, followed by reflective observation, abstract conceptualisation and active experimentation (Kolb 1984).

The problem with traditional problem-based learning and teaching, based on active and experiential learning, such as case competitions and class-based simulations in real estate education, is that they are resource-intensive and rely heavily on the selection and performance of the industry participant. Without adequate guidance and resourcing, there is a risk that the authenticity and depth of inquiry may be compromised. For example, the use of industry-standard tools may align with stakeholder expectations for practice-ready graduates (API 2021; Azasu & Gibler 2016), but it risks shifting the focus of the learning from critical engagement with underlying valuation principles to the mechanical application of software. Palm and Pauli (2018) caution against this shift, noting the potential for an overemphasis on practicalities at the expense of theoretical depth.

This distinction between training and education is echoed by Wilkinson, Halvitigala, and Antoniades (2018), who argue that while training equips students with the ability to use specific technologies, education should cultivate a deeper understanding of foundational theories and the capacity for critical inquiry and lifelong learning.

Similarly, Mooya (2015), in his critique of South African real estate curricula, challenges the prescriptive application of vocational training models to valuation education. He argues that aligning specific property types with fixed valuation methods undermines the development of broad conceptual knowledge, knowledge that should be the hallmark of a university education.

The hallmark, or even role, of university education more broadly continues to be contested in the 2024 Contextualising Horizon Report, with Arumugam, Warnakula and Todd, in Porter, Campbell, Jones, Logan-Fleming and Sankey (2024), noting, 'even though university degrees are still considered important, society is beginning to question the value of these qualifications as a key to securing a job' (p. 3). They discuss skills development in higher education, highlighting the potential corporate social responsibility of institutions to be purpose-driven and to incorporate more work-integrated learning (WIL).

The same report has disruption as the recurring theme. Porter et al. (2024) refer to disruption in the sense of a break from the status quo for higher education, with calls for work-ready graduates, skills-based education and new models and learning pathways expected. Disruption in the Australasian region, from economic challenges, pandemics, earthquakes and increasingly volatile weather events (Porter et al. 2024), makes accessing and sustaining these new education models and learning pathways more critical.

Accessing and sequencing active and experiential learning

The necessity to physically distance during the COVID-19 pandemic changed the university delivery models, making education more accessible. Providers and students moved to online materials, not dissimilar to massively open online courses (MOOCs), the 'educational revolution to override borders, race, gender, class and income' (Emanuel, Christensen, Steinmetz, Alcorn, Bennett & Woods 2013, p. 342). That said, more active and deeper learning approaches, as discussed prior, are resource-intensive and not suited to current MOOCs and online delivery (Boyd 2015). For these reasons, and others, universities revisited blended and asynchronous learning, previously viewed as a mechanism to enhance face-to-face delivery (Poon 2014), in the development of pedagogical models as a response to COVID-19 (Petronzi & Petronzi 2020).

An example of a structured blended learning framework is the Online and Campus (OaC) Model developed by Petronzi and Petronzi (2020). Originally conceived in response to the disruptions caused by the coronavirus pandemic, the model is proposed as a sustainable and progressive approach to learning and teaching in higher education. The model, incorporating suggested examples, a summary of student experience and tutor roles and responsibilities, is presented in Figure 9.1, Online and Campus (OaC) Model.

A depiction of the Online and Campus (OaC) modei with a suggested route and summary of tutor roles and student experience for each stage of the learning process,

	Suggested Example	Opportunities and Roles
[1] Introduce Knowledge ↓	Asynchronous:	**Student Opportunities/Role:** [1] Flexible and accessible engagement with knowledge and theory. [2] Ability to work at own pace and access different representations of the same tyt [3] Develop academic and graduate employability skills.
		Student Opportunities/Role [1] Representation of content to make accesible. [2] Embedding of teaching tasks. [3] Formative opportunities through tasks [3] Opportunity to provide individualised feedtack bls.
[2] Embed/ Conceptualise ↓	Synchronous:	**Student Experience:** [1] Asynchronoi/knowledge transmission [2] Direct tutor contact. [3] Less social pressure than campus/more anonymity if supported correctly can equal higher engagement and more confidence to respond:
[3] Apply/ Challenge ↓	**Campus:**	**StudentOpportunities/Role** [1] Problern-based learning [2] Greater questioning both bnench per and tutor supporort. [3] Self-reflection/self-assessmoent opportunity. [3] Structured self-assessment and team buidliding. [5] Social opportunities and team butla [1] Formalised review of taught knowledge adapting task to group up needs appro [2] Practical opportunities and teaching (where required).

Figure 9.1 Online and Campus (OaC) model (Petronzi & Petronzi 2020, p. 502).

The Petronzi and Petronzi (2020) approach commences with a self-directed, asynchronous phase focused on acquiring foundational knowledge and theoretical concepts. The next stage is synchronous, where students engage with the material to deepen understanding and conceptual integration. The final phase involves on-campus learning, where students apply, test and critically evaluate their knowledge through problem-based learning activities.

This structure presents promising pathways and justification for the blending and alignment of synchronous and asynchronous learning. That said, it does not fit neatly with disciplines and specialisations such as real estate. In real estate, specialised property units are often squeezed into the central period of a student's education programme to 'accommodate generic socialisation units in the first year and non-property related skills, soft skills and industry engagement in the final year' (Armitage 2011, in Parker 2012, p. 4).

A new real estate education framework

For real estate education, the Petronzi and Petronzi model may be adapted, as demonstrated by Boyd (2022), to restructure a valuation modelling module as the first asynchronous individual learning activity. In this restructure, the initial asynchronous phase is applied in nature, requiring valuation model construction, forming a foundation for subsequent learning. This is followed by a synchronous stage, where students analyse and evaluate various valuation methods, deepening their conceptual understanding. The final phase consists of an active learning workshop, delivered either on campus or in a virtual setting, that incorporates elements of problem-based learning to share and problem-solve. This structure is illustrated in Figure 9.2, Valuation Modelling Unit Structure.

The structure of the valuation modelling unit incorporates a novel framework known as apply, evaluate, and solve (AES). This model supports flexible, self-paced learning, allowing students to engage with foundational content and develop a working valuation model at their own pace. Learners may choose to exit the module upon achieving an introductory level of competence, with the potential for this achievement to contribute towards a lower-level qualification, certificate, or professional recognition through non-traditional credit pathways (McGrath et al. 2020).

Students who continue beyond the initial phase enter a learning community, bringing with them prior experience in valuation model development. Within this community, learners engage in deeper analytical tasks, evaluating valuation approaches, reflecting on their own models, and critically examining the assumptions and inputs that underpin valuation practices. The final stage of the unit involves an intensive workshop, where students apply advanced cognitive skills to collaboratively seek answers to complex problems or simulated critical scenarios (Boyd 2022).

In bringing forward the practical, or applied, part of the valuation modelling, the sequencing of this unit aligns more closely with the progression outlined in the Australian Qualifications Framework (AQF). Specifically, the initial 'apply' phase corresponds with AQF Level 5, with graduates having '… specialised knowledge and skills for skilled/paraprofessional work and/or further learning' and '… [applying methods and technologies to: analyse information to complete a range of activities'. (AQF 2021)

The subsequent stages in the AES framework more closely resemble the criteria prescribed in the higher undergraduate and postgraduate university levels, 6 through 8. They are also well suited to case-based learning and active learning, encouraging students to adopt deeper learning strategies (Marton & Saljo 1976a, 1976b; Biggs 1993; Boyd 2005; Biggs & Tang 2009; McGrath et al. 2020).

While the AES framework provides a structured and progressive approach to learning, its implementation is not without potential challenges. As Azasu and Gibler (2016) caution, designing a unit that attempts to cover an overly broad scope of valuation modelling theory may dilute pedagogical effectiveness, potentially overwhelming students and hindering deep understanding. Furthermore, without clear instructional guidance, there is a risk that the framework could be misinterpreted. Specifically, it may be mistakenly viewed as a reversal of the challenge-driven learning model advocated by McGrath et al. (2020), where students are expected to apply knowledge before it has been adequately developed.

Stage	Learning activities and material	Student role	Instructor role	Assessable deliverables
1. Introduce knowledge and apply - develop a working valuation model. [Apply]	Valuation modelling module. As online asynchronous individual learning.	Engages with knowledge theory and works at own pace to develop an understanding of valuation approaches while developing a prescribed valuation model.	Facilitation through providing formative individualised feedback. Provides pathways for the student to access learning resources customised to their journey.	Valuation model Answered quiz
2. Analyse and evaluate valuation methods and the inputs that inform their functioning. [Evaluate]	Experiential learning embedded in a learning management system. Synchronous including scheduled virtual meetings and discussion tasks. Still focused on individual learning.	Sources and interprets specific market information, analyses sales, and confirms assumptions. Justifies benchmarks and assumptions and debates theoretical underpinnings of the valuation methodologies adopted and considered.	Facilitates formation of study groups and demonstrates pathways to obtain specific market information, sales details, and confirm assumptions. Provides pathways for the student to access learning resources customised to their journey.	Refined and specialised valuation model incorporating justification for inputs. Personal reflections.
3. Design, develop and share a real solution for a set problem or simulated critical event. [Solve]	Active and intensive learning workshop embedding aspects of problem-based learning. Set on campus or in a virtual meeting. Synchronous group learning.	Cooperatively analyses and interprets the problem or simulated critical event. Plans a pathway and coordinates activities to formulate a solution. Models the financial impact of the solution. Develops the narrative and presents the solution.	Provides problem instruction as a simulated client. Observes operation of team and coordinates the assessment panel.	Refined and/or redesigned valuation model. Face-to-face or multimedia sharing of the proposed solution.

Figure 9.2 Valuation modelling unit structure (apply, evaluate, and solve) (Petronzi & Petronzi (2020), Boyd (2022) & Author).

Another limitation of this framework lies in its reliance on self-directed, asynchronous learning during the initial 'apply' phase. This stage demands the adoption of new strategies, tools and approaches to optimise resource use and simulate the guidance typically provided by an educator or coach. While serious games (Boyd 2015) and collaborative activities can enhance engagement and support the achievement of learning outcomes, a comprehensive suite of educational technologies may be necessary to compensate for the lack of early-stage mentoring and support.

Educational technologies, including AI, seek to foster levels of engagement comparable to face-to-face learning while providing the accessibility and flexibility associated with online modalities. Specifically, the goal of AI in education (AIED) for kindergarten to year 12 (final year of school) is to develop systems that enable personalised, flexible and engaging learning and to automate mundane teaching tasks (Southgate et al. 2018).

Before applying educational technologies to the real estate disciplines in higher education, it is important to consider how related technologies are shaping the industry and future professionals. This next section specifically focuses on technology in valuation, or appraisal, with explicit consideration of robots and drones, computer vision and LiDAR and AI.

Advances in technology and valuation

The practice of real estate valuation is undergoing a significant transformation as emerging technologies and systems enable greater automation, enhanced analytical capabilities and more sophisticated forms of intervention. The advent of Industry 4.0, or the fourth industrial revolution, has introduced more than just smart factories and products. As Starr, Saginor, and Worzala (2020) explain, it has provided adaptable smart processes and systems driven by advances in artificial intelligence, cooperative robotics and adaptive automation control.

These innovations are reshaping the real estate industry (Ullah, Sepasgozar, Thaheem & Al-Turjman 2021b; Starr et al. 2020; Porter, Fields, Landau-Ward, Rogers, Sadowski, Maalsen, Kitchin, Dawkins, Young & Bates 2019; Ullah, Sepasgozar & Wang 2018) and the practice of real estate valuation (Abidoye & Chan 2017a, 2017b; Valier 2020; Boyd, Small & Boyd 2024). AI-driven interventions are changing the way we value real estate, presenting the contemporary valuer with a new suite of skills to pursue (Boyd et al. 2024).

The conceptual research by Boyd et al. (2024) sought to better understand the new suite of valuer skills through relationship mapping. The mapping considered the capabilities of AI and how those could augment the valuation process. In their findings, they spoke of the physical real estate inspection, required to address the land and improvements component of a valuation, as having the greatest number of AI capabilities to explore with,

> Learning robots or drones with computer vision could enhance the current inspection process accessing areas where the human cannot, capturing and analysing far more unique characteristics than the human valuer can.
>
> (Boyd et al. 2024, p. 14)

In the API (2023) survey, virtual tours are said to be becoming commonplace, as

> end users note the advantage of time, cost and carbon emissions savings [and over time] there may also be an opportunity for headsets, and even virtual reality glasses to be used to aid property inspections.
>
> (API 2023, p. 44)

The emerging opportunities for robotics, and drones, imaging, including LiDAR, in real estate are presented in research building on systematic reviews, such as the review of smart real estate technology by Ullah et al. (2018).

Robots and drones

Ullah et al. (2018) and Sepasgozar, Davis and Loosemore (2018) found that while robots were applied in medicine to enable people with physical disabilities to conduct complex and precise operations, they are increasingly used in construction and real estate. Robots can assist or facilitate real estate inspections and, when equipped with other technologies, can transport clients and customers to properties and develop 3D renderings and models of internal spaces (Ulla et al. 2018).

Originally developed for military use, drones, also known as unmanned aerial vehicles (UAVs), have been increasingly adopted in real estate management for data collection through aerial photography and 3D imaging (Sepasgozar et al. 2018; Ullah et al. 2018). These technologies enable real estate professionals to capture and present perspectives of properties that would otherwise be inaccessible or prohibitively expensive to obtain using traditional methods such as helicopters or aircraft. Prospective buyers can remotely assess property views, proximity to neighbouring structures and surrounding amenities. According to Ullah et al. (2018), Luppicini and So (2016), and Newell (2017), the integration of drone technology has significantly reduced inspection times and contributed to improved sales outcomes.

Sepasgozar et al. (2018) and Ullah et al. (2018) highlight the emergence of advanced robotic and drone technologies equipped with high-resolution cameras and both aerial and wall-climbing capabilities. These innovations offer novel perspectives for viewing entire properties and their surrounding neighbourhoods. The resulting imagery provides consumers with broader, more comprehensive visual angles, enhancing their ability to assess a property remotely and make informed decisions.

Drone-captured visuals deliver valuable spatial data, including sun orientation, surrounding green infrastructure, and proximity to essential amenities such as schools and town centres (Ullah et al. 2018). Within the property, high-quality, zoomable interior images reveal fine architectural and design details from otherwise inaccessible viewpoints, enhancing the consumer's understanding of the space and place (Ullah et al. 2018).

Computer vision and LiDAR

Beyond high-resolution imaging, 3D scanning technologies, such as LiDAR, are proving to be transformative in both real estate and construction sectors (Ullah et al. 2018). Scanners such as LiDAR are used to generate precise as-built digital models, which can be employed to update architectural drawings, monitor maintenance, and track repairs (Sepasgozar, Forsythe & Shirowzhan 2016; Takin, Peng, Sepasgozar, Ebrahimi 2017; Ulla et al. 2018). Takin et al. (2017) suggest 3D scanners as a convenient and low-cost approach to inspect residential property. These scanners are not only approximately 50 per cent more cost-effective than traditional inspection methods, but they also enable construction managers to efficiently update project data and revise architectural drawings using dynamic, multidimensional models (Takin et al. 2017; Ulla et al. 2018). For real estate consumers, 3D-scanned images and digital drawings offer a realistic representation of a property, particularly valuable when in-person inspections are not feasible. This enhanced visualisation helps buyers better understand spatial layouts and design features, supporting more informed decision-making. Additionally, such models allow prospective owners to plan renovations, modifications, or extensions with greater confidence (Ulla et al. 2018).

The emerging technologies are changing the way we value and inspect real estate. Some valuers may choose to physically inspect to touch, feel, smell or gain a nuanced sense of place and space. That said, existing products such as Elios 3 (Flyability 2024c), LiDAR and multi-sensor-enabled drones are arguably better than the human equivalent at mapping and inspecting complex properties and infrastructure. Young et al. (2024) used the compact, collision-tolerant indoor inspection drone with an integrated lightweight LiDAR system, using simultaneous localisation and mapping algorithms, and a dosimeter (radiation) payload as a safe and effective approach for collecting the 3D spatial and radiation survey data from inside a very high radiation storage vault. According to Young et al. (2024), the drone, as depicted in Figure 9.3, performed in the very high radiation and high heat nuclear waste storage vault for

> two flights during which the drone's LiDAR unit was used to collect 3D spatial data that completely characterised the location of all obstructions … The drone's carbon [fibre] cage and semi-autonomous stability systems were key features for enabling the drone to successfully [manoeuvre] in the obstructed area.
>
> (p. 11)

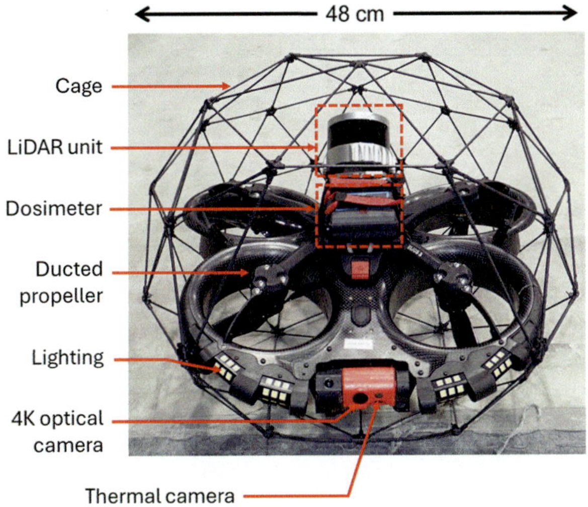

Figure 9.3 Photograph of the Elios 3 prototype drone configuration with integrated LiDAR and dosimeter (Young et al. 2024).

With respect to real estate inspections and mapping, Flyability has a series of case studies that demonstrate how the Elios 3 can effectively map or digitise (Flyability 2024b & Flyability 2024d) and inspect (Flyability 2024a) complex properties without the need for a physical human presence. It is important to note that the product, digital resources, data and videos associated with these mapping exercises and inspections provide far greater understanding of the asset than provided from a physical real estate inspection. To clients and end users, this may be regarded as a benefit or service for which a premium may be provided. Conversely, a valuer or appraiser may be apprehensive about providing such detailed information, especially if it has not been fully discussed or addressed in the assessment of value. For example, the inspection process may identify structural problems that are beyond the experience of the valuer to identify or quantify.

These more detailed mappings and inspections may require a different skill set and ability. Rather than being physically mobile, the new real estate valuer may be better served by developing understanding, knowledge and skills that are more diverse and aligned to partnering with emerging technologies. Put another way, real estate employers may be better served seeking a more diverse and inclusive workforce where disabilities and disorders are no longer barriers.

Emerging inclusive workforce

Persons with disabilities include 'those who have long-term physical, mental, intellectual or sensory impairments which in interaction with various barriers

Figure 9.4 Photograph of LiDAR data capture with the Elios 3 on Alcatraz (Flyability 2024d).

Source: nexxis.com.au

Figure 9.5 Photograph of the Elios 3 conducting a visual inspection of key parts of water rides and a roller coaster (Flyability 2024a).

Source: nexxis.com.au

Figure 9.6 Photograph of schematic design of a plant section used for engineering layout and spatial design of new plant structure assisted by the Elios 3 (Flyability 2024b).

Source: nexxis.com.au

may hinder their full and effective participation in society on an equal basis with others' (UN 2006, np). The Australian *Disability Discrimination Act* 1992 prescribes a more explicit reference to a disability in relation to a person as:

- loss of bodily or mental functions; or
- loss of part of the body; or
- presence of organisms causing, or capable of causing disease or illness; or
- malfunction, malformation or disfigurement of a part of the person's body; or
- disorder or malfunction that results in the person learning differently; or
- disorder, illness or disease that affects a person's thought processes, perception of reality, emotions or judgement or that results in disturbed behaviour. (Disability Discrimination Act 1992)

The physical and mobility-related disabilities remain a barrier to real estate valuation, as the Australian Property Institute requires valuers to visit a property or tangible asset (API 2024, 11.2). That said, the interventions and technologies previously addressed may provide the less mobile with mobility and even facilitate virtual inspections with the API survey findings. Virtual tours are becoming commonplace (API 2023). The disabilities that refer to mental functions and judgements are likely to remain a barrier to registration as a real estate valuer as the industry, institutes and registration bodies seek valuers of 'good fame and character and … fit and proper person[s]' (Valuers Registration Act 1992, S.30a). The definition for a person of good fame and character and a fit and proper person is not explicitly addressed in the *Act* or guidelines, which leaves the decision on suitability to the appointed board. In a practical sense, a person exhibiting 'disturbed behaviour' (Disability Discrimination Act 1992) could have a related police record and be deemed as not meeting the board's fit and proper test.

A level of discrimination, with respect to persons exhibiting thought processes, perception of reality, emotions or judgement or that results in disturbed behaviour, may remain appropriate to protect the integrity of the valuation decision-making process. But other barriers, such as the requirement for physical real estate inspections and the associated inherent requirements for mobility, appear unsustainable and not reflective of the future role of a real estate valuer. Educators should bear this in mind and do what they can to provide reasonable adjustments to enable valuable work experience for remote, disadvantaged and disabled high school students. AI in education (AIED) can assist with these adjustments and provide improved equity in education with reduced educational disparities, improved accessibility to education and reduced costs of education (Păvăloaia & Necula 2023).

Artificial intelligence in education

AI in education (AIED) has emerged as a specialised interdisciplinary field focused on enhancing learning and instruction through AI technologies (Southgate et al. 2018). Jimenez and Boser (2021) note that AI can help students learn more efficiently by identifying issues that may not be visible to teachers, allowing early intervention. Southgate et al. (2018) propose that AIED has the potential to reduce achievement gaps by supporting personalised, flexible and engaging learning experiences, while also automating tasks such as assessment and feedback.

Key developments in AIED include intelligent tutoring systems, virtual pedagogical agents, embodied AI robots, and smart classrooms. Intelligent tutoring systems simulate one-on-one instruction, though their effectiveness may vary across learners. Pedagogical agents, virtual avatars, enhance engagement by adding social and emotional dimensions to digital learning. Smart learning environments integrate wireless devices, sensors and platforms to adapt both the physical and instructional settings in real time.

AI also supports adaptive learning by adjusting content difficulty, task types or interfaces to suit individual needs. In higher education, AI-powered learning environments generate large datasets on student engagement and performance. These are analysed through learning analytics and educational data mining, enabling content to be tailored and updated using deep learning techniques (Southgate et al. 2018). However, this level of personalisation raises important ethical concerns, particularly regarding data privacy and student rights.

Drawbacks of AI in real estate

Despite the promise of artificial intelligence, its adoption in real estate and education raises significant risks and socio-political concerns. As Landau-Ward and Porter (in Porter et al. 2019) argue, digital technologies are not neutral tools; rather, they are embedded within broader social and political power structures, echoing Kitchin's (2014) assertion that digital disruptions reshape, rather than simply replace, analogue processes. Braesemann and Baum (2020) similarly caution that while PropTech innovations offer substantial efficiency gains, they also risk transforming the real estate market into data-driven, oligopolistic systems dominated by a few powerful firms. To counter this, they stress the importance of real estate users and owners recognising the value of the data they generate.

These views reflect broader concerns about the power dynamics introduced by AI. Wagner (2021) warns that automation and AI augmentation can erode humanistic perspectives in leadership, reducing ethical oversight and increasing the potential for misuse. He describes AI as a powerful managerial tool, capable of both positive and harmful outcomes, depending on how it is controlled and applied.

The darker implications of AI are exemplified by the development of 'Norman', an AI system trained on specific data to produce psychopathic interpretations of inkblots, as reported by MIT Media Lab and Scalable Cooperation (MIT 2021). This case underscores the dangers of training AI in uncontrolled or biased environments. Kumar et al. (2019) emphasise the need for rigorous oversight in AI development, warning that poorly designed systems can be manipulated to pursue harmful objectives under the guise of fulfilling programmed goals. These examples highlight the need for ethical, transparent and human-centred approaches to AI design, development and application.

Human oversight in recommender systems and nudging

Human oversight remains critical in the use of AI in education, particularly with systems like recommender algorithms and emotive 'nudging' technologies. While AI can personalise learning pathways, systems designed to guide student choices, similar to those used by platforms like Netflix, can inherit biases that may unintentionally infringe on students' rights (Jimenez & Boser 2021). Rather than aiming to eliminate bias entirely, Jimenez and Boser (2021) argue that policymakers must determine what level and type of bias is acceptable and ensure it does not disproportionately affect students based on race, income, disability or language background.

Affective computing and nudging systems, which influence learners' emotional states, raise ethical concerns about autonomy and informed decision-making (Southgate et al. 2018). Educators and researchers have expressed unease about partnering with AI systems that exert influence without empathy (Boyd 2021). He warns that as educators focus on training AI, they may inadvertently reduce opportunities to mentor human learners, particularly those from disadvantaged or remote backgrounds.

Training AI is likely to become an additional skill expected of future valuers. If this expectation grows, it may be necessary to reconsider the role of educators and placement providers in developing skills that AI may eventually assume. In response to these shifts, Boyd et al. (2024) recommend that education providers prioritise the development of:

- Humanistic decision-making
- Multidimensional information mastery (including numeracy and financial analysis)
- Leadership and management skills
- Time management
- Effective communication (but not specifically report writing, oral or written)

In a broader higher education sense Gribble, Gander, Tsatsaronis and Birt, in Potter et al. (2024), make similar assertions around the need for humanistic decision-making and information mastery with the warning, 'educators and the general public will need to consider what they trust and when and to develop critical thinking skills to reduce a trend towards "blind trust" as AI tools continue to proliferate' (p. 17).

In real estate, skills and attributes that lead towards ethical decision-making will become more sought-after with further responsibility placed on fewer people.

This research builds on those skills and attributes as it shares the design of simulations and practical activities, using available technologies. The simulations and their proposed incorporation in the high school curriculum are demonstrated as contributions.

Real estate simulations

Case-based studies and experiential learning have a rich history in real estate education, with case studies based on current real estate problems of current executives encouraged in teaching. In discussing the future of real estate education, McGrath et al. (2020) encourage the adoption of experiential learning approaches and methods such as challenge-driven learning. The challenge-driven problem-solving enables students to use knowledge gained in the classroom via more traditional approaches and apply it to the real-life challenges that will require them to create, invent and develop new ideas, approaches and solutions.

This research adapts case-based studies, or simulations, designed for learning the financial analysis inherent in real estate valuation asynchronously, with a sequencing framework and AIED interventions to enable valuation work experience for remote, disadvantaged and disabled high school students. As a technology-enhanced WIL placement, Dickson-Deane, Heggart and Logan-Fleming, in Porter et al. (2024), the pursuit is one way to address future employability needs. In a broad university context, they conclude that 'building WIL opportunities and experiences is the bridge that many learners need to prepare for the real world' (p. 22).

This artefact, or technology-enhanced WIL package, comprises four real estate simulations, or case studies. The simulations provide a dual purpose, being educational and promotional. The fundamental design and focus on skills-based learning in the simulations have a constructivist base with phasing that encourages early application (Boyd 2022). From a promotional sense, this simulation package can provide a taste of real estate careers for school students undertaking work sampling placements.

The simulations

The simulations are asynchronous with the potential to embed learning analytics and education mining to enable the content and learning to be tailored and updated as markets and practices change. With appropriate human oversight, AI recommender systems may be purposed to suggest academic pathways for the student or candidate to pursue.

The simulations present the journey of Jamiee, an assistant valuer, as she explores 'a day in the life' of a series of real estate professionals, including valuers, an asset manager and a developer. The students, or candidates, virtually join Jamiee on real estate inspections and other activities, then complete a series of tasks to assist the host real estate professional. The simulations and activities are detailed in Table 9.2: Experience real estate simulations.

In addition to applying problem-solving and analytical skills sought after in real estate and finance, the simulations showcase real estate technologies used, including robots, drones, computer imaging and LiDAR. There is the capacity for the simulations to be partnered with structured formal activities and training, such as drone courses and computer image training. Assuming appropriate certification is complied with, university libraries may distribute drones and enabling technologies in a similar manner to books and physical learning resources.

Besides physical resources such as drones, robots and computer imaging tools, this is a project designed to complement existing training programmes. Rather than modules set on specific knowledge fields, the simulations are based on the practical and hands-on activities undertaken on a day-to-day basis by

Table 9.2 Experience real estate simulations (Author)

Simulation	*Brief*	*Tasks*	*AI capabilities*	*AIED*
1 House Valuer	Work as a valuation cadet in New South Wales' northern beaches. Join Pat inspecting and measuring a home and then compare it with others to assess market value.	1.1 Measure the area of a house (60–90 min) 1.2 Calculate the net income (20–30 min) 1.3 Apply comparable sales to determine value (60–90 mins)	Drones, robots and computer imaging enabled by natural-language text and speech understanding and generation, virtual agents or conversational interfaces, and recommender systems	Adaptive learningLearning analytics and education mining content creation and learning may be tailored and updated as markets and practices change
2 Infrastructure Valuer	Join the world of Hi-Viz and PPE as you assist Jeff to value a customs facility at the Fremantle port. You will need to look beyond the blue water and breweries as you assess the value for financial reporting purposes.	2.1 Estimate quantity and cost 2.2 Determine replacement cost and assess the insurable value 2.3 Apply depreciation to assess the depreciated replacement cost	Drones, robots and computer imaging enabled by natural-language text and speech understanding and generation, virtual agents or conversational interfaces, and recommender systems	Adaptive learningLearning analytics and education mining content creation and learning may be tailored and updated as markets and practices change

3	Asset Manager	Assist Jane with the financial analysis of a shopping centre in Noosa. Perform investment analysis and evaluate the viability of installing a new solar array.	3.1 3.2 3.3	Populate a tenancy schedule and establish an outgoings budget Apply Discounted Cash Flow analysis Evaluate scenarios in financial analysis and decision-making.	AI-enabled financial analysis tools with natural-language text and speech understanding and generation, virtual agents or conversational interfaces, and recommender systems	Adaptive learningLearning analytics and education mining content creation and learning may be tailored and updated as markets and practices change
4	Developer	Experience working as a developer's assistant on Queensland's Gold Coast. Join Linda as she inspects a development-ready property overlooking Surfers Paradise beach and help her explore development options to find the optimal use of the land and its expected sale price.	4.1 4.2 4.3	Identify the property and assess potential uses (60–90 mins) Analyse sales to determine best use (60–90 mins) Apply residual analysis to assess land value (60–90 mins)	Drones, robots and computer imaging enabled by natural-language text and speech understanding and generation, virtual agents or conversational interfaces, and recommender systems	Adaptive learningLearning analytics and education mining content creation and learning may be tailored and updated as markets and practices change

real estate professionals. They do not provide a full education offer but rather present a suite of activities that simulate what real estate professionals do. The activities can form an e-portfolio for later registration and membership. But, in a higher education context, the focus on the tasks should be on the assessment for learning, not of learning. Subsequent university units or modules, more focused on evaluation and solving or planning, would form a more rigorous base for comparative grade allocations.

The simulations have the potential to enhance learning and promote real estate as a career option, as well as raise the profile of real estate degrees for Australian universities. For high school students and aspiring real estate professionals, the embedded activities may form part of an e-portfolio to evidence practical experience. With appropriate funding and dissemination, the simulations can provide a critical link to high school students across Australia at a stage where they are career planning.

While there may be sound grounds for the adoption of simulations as work sampling placements, in Australian high schools, there are existing barriers. The conditions of existing work experience placements are framed around a physical workspace and synchronous learning. For example, the *Education (Work Experience) Act 1996* applies the condition that 'the work experience provider must not provide work experience to the student at a time other than during the ordinary working hours of the place where the work experience is provided' (S12(f)). The current work experience agreement, provided in the Department of Education's Procedure (2022), places further conditions requiring the employer, or work experience provider, to provide 'direct supervision at all times while undertaking work-related activities', allocate a 'workplace supervisor to the student', and 'ensuring the permitted number of students accepted for work experience does not exceed the number of full-time employees' (Department of Education 2022, np).

Engaging with intelligent systems

Engaging with artificial systems can take a variety of forms. In discussing Artificial Engagement in Instructional Design (AEID), we have identified opportunities and threats inherent in AI assuming the role of an educator or teacher, a matter addressed further in Boyd (2021, 2023), as is the role of the human in teaching AI. This presents the question: how can we humanely engage with intelligent systems?

The design, implementation, and governance of AI in education involve complex ethical and human rights considerations (Southgate et al. 2018; Jimenez & Boser 2021). These concerns are explored in the IEEE's Ethically Aligned Design, which outlines principles to ensure autonomous systems remain human-centric and aligned with societal values (IEEE 2017). Similarly, the Australian Human Rights Commission's technical paper Addressing the

Table 9.3 Consumer Policy Research Centre consumer outcomes (Lattimore et al. 2020, p. 57)

Accessibility	Markets are inclusive, and all consumers have the right to access this technology and its application on an equal basis with others.
Accountability	Consumers have a clear route for seeking explanations and accessing appropriate redress from a responsible party if things go wrong.
Agency	Consumers are empowered to exercise autonomy and freedom of choice in their interactions with technologies such as AI systems and the use of their personal data.
Transparency	People are made aware when they are the subject of a decision-making process that uses an AI system.
Understandability and explainability	Individuals subject to these decisions are entitled to a meaningful, comprehensible explanation of the AI system and its decision-making process.
Sustainability	Long-term implications of technology on consumers are considered and addressed throughout design and implementation.

Problem of Algorithmic Bias presents key questions and six consumer-focused outcomes to guide responsible AI and data use in business (Lattimore et al. 2020). The Lattimore et al. (2020) consumer outcomes are presented in Table 9.3: Consumer Policy Research Centre consumer outcomes.

In the context of AIED, Southgate et al. (2018) identify five foundational pillars for ethical AI: awareness, explainability, fairness, transparency and accountability. These principles guide the ethical design, implementation and governance of AI interventions in education, as illustrated in Table 9.4: Applying the Five Pillars of AI Ethics to School Education.

Boyd (2021) designed the Intelligent System Partnering Framework (ISPF) for the practice of valuation. This framework draws on lessons from AI adoption in the sector and incorporates practical strategies, including those proposed by Sheehy et al. (2020) for managing challenging behaviours in professional environments.

Together, the Australian Human Rights Commission's consumer outcomes, Southgate et al.'s (2018) five ethical pillars for AIED, and Boyd's ISPF provide a foundational structure for guiding AI use in real estate education. Through synthesising these elements, a new framework, Artificial Intelligence in Real Estate Education (AIREE), has been developed, offering a model for ethical and effective AI integration in teaching and learning.

Table 9.4 Applying the five pillars of AI ethics to school education (Southgate et al. 2018, pp. 48–49)

	Design	*Implementation*	*Governance*
Awareness	How have the manufacturers of the system engaged with the education stakeholders to raise awareness of AI, its limitations, potential and risks?	Have students and parents/caregivers been made aware of the type of data harvesting and sharing arrangements required by the system?	Is there a rigorous process for seeking parental consent and student assent before systems are deployed?
Explainability	Is the system designed to explain to students, parents and teachers its purpose, process, decisions and outcomes in an accessible way?	What opportunities, approaches and public forums are available for students and parents to explore, explain and share information and experiences of AI in schooling?	Do policymakers, procurement officers and school leaders have access to appropriate independent technical expertise to explain and advise on AI systems?
Fairness	Has the issue of potential bias in the design of the system been proactively addressed and documented?	How will the school address potential inequalities in an AI world? Does the system use autonomous experimentation, and could this create an unfair burden on students and teachers? Does the AI system introduce unjust and punitive types and levels of surveillance on students and teachers?	What procedures and policies are there to ensure that AI systems positively address rather than exacerbate inequity, discrimination and prejudice in education? What evidence is there that an AI system can be used to address equity concerns in schools?

Transparency	Is the system designed and implemented for traceability, verifiability, non-deception, honesty and intelligibility?	Can students, teachers, parents and the community inspect and have opportunities to respond to AI systems training and decision-making in ways that are intelligible or authentically empowering to them?	How will those in governance or procurement positions ensure genuine traceability, verifiability, non-deception and honesty and intelligibility of AI systems prior to purchase and during implementation? How will transparency be operationalised if harm occurs?
Accountability	Have the designer and vendor of an AI system clearly articulated their responsibilities to the ethical use of AI? What systems do they have to ensure ethical accountability?	Who is accountable for the procurement of ethical AI? Is there a school- and system-wide procedure for reporting and responding to AI harm? Do all stakeholders in the school community know about and how to access the above procedure?	What protocols are in place to prevent and respond to harm? What early warning systems are there that harm may be occurring that can trigger action?

Framework design for humane AIREE

The merged framework represents an evolution of Boyd's (2021) Intelligent Systems Partnering Framework (ISPF), adapted to incorporate ethical and governance considerations specific to AI in education. In the initial Identification (I) stage, any proposed system must satisfy all six consumer outcomes outlined by Lattimore et al. (2020): Accessibility, Accountability, Agency, Transparency, Understandability and Explainability and Sustainability.

The Due Diligence (DD) stage introduces a new requirement, compliance with the AIREE due diligence process. This process closely aligns with the governance questions derived from Southgate et al.'s (2018) AIED ethical pillars and focuses on five key questions:

1 Is there a rigorous process for obtaining user consent?
2 Do educators have access to independent technical expertise?
3 Does the system promote fairness?
4 Is the system transparent?
5 Are there protocols in place to prevent and respond to harm?

The final question is operationalised in the Management (M) stage. If the system fails to perform as intended, protocols must be activated to mitigate harm and, if necessary, discontinue its use.

This integrated model is illustrated in Table 9.5: AIREE Partnering Framework.

The AIREE framework offers real estate educators a structured approach to engaging with AI in teaching and learning. It comprises five stages:

1 Identification – Evaluates the AI system's potential impact on human rights, aligning with ethical standards.
2 Control categorisation – Assesses the level of autonomy granted to the AI, which determines the appropriate pathway through the remaining stages.
3 Viability and suitability – Considers cost-benefit factors, including the integration of human oversight as a critical component of system viability.
4 Due Diligence – Focuses on governance, drawing on the ethical pillars of AIED (Southgate et al. 2018) to ensure responsible implementation.
5 Management – Monitors system performance and determines whether further evaluation or corrective action is required.

This framework is designed to support informed, ethical and adaptive integration of AI into real estate education.

Table 9.5 AIREE Partnering Framework (Author and Boyd 2021)

Identification (I)		Control (C)		Viability and suitability (V)		Due Diligence (D)		Management (M)	
I1. **Intelligent system or information sharer**? Consider, is the proposed intervention an intelligent system or does the system have the potential to share your data or analysis? If so, does it meet all the Human Rights consumer outcomes of: 1. Accessibility 2. Accountability 3. Agency 4. Transparency 5. Understandability and explainability 6. Sustainability	Yes (C1)	C1. Do you have **full control** of the intervention? Consider, has it been developed internally or purchased for exclusive use?	Yes (V1) No (C2)	V1. Do the projected benefits outweigh the anticipated costs and need for periodic **human monitoring**?	Yes (D1) No (I1)	D1. Does the intervention pass **AIREE due diligence**, being? 1. Is there a rigorous process for seeking user consent? 2. Do educators have access to independent technical expertise? 3. Does the system promote fairness? 4. Is the system transparent? 5. Are there protocols in place to respond and prevent harm? 6. Can the intervention be **acquired retaining full control**? and 7. Can you **set the performance benchmarks**?	Yes (M1) No (I1)	M1. Does the system **perform as prescribed**?	Yes (continue)
	No (alt. acquisition process)	C2. Do you **share control** of the intervention? Consider, can you set the role and determine what the system can and cannot do?	Yes (V2) No (C3)	V2. Do the projected benefits outweigh the anticipated costs and need for **sustained human monitoring**?	Yes (D2) No (I1)	D2. Does the intervention pass **AIREE due diligence**? Can the intervention be **acquired on terms that mirror an employment agreement**? And can you **set performance benchmarks**?	Yes (M2) No (I1)	M2. Does the system **perform as prescribed**? Consider, **has the role changed** and are the same system controls in place?	No (initiate protocols to respond to potential harm and discontinue – I1)
		C3. Does a system or **another person control** the intervention? Consider, does another person or system set the terms and conditions for use?	Yes (V3) No (I1)	V3. Do the projected benefits outweigh the anticipated costs, need for sustained **human monitoring, and opportunity cost** of intellectual property?	Yes (D3) No (I1)	D3. Does the intervention pass **AIREE due diligence**? And can you **set performance benchmarks**?	Yes (M3) No (I1)	M3. Does the system **perform as prescribed**? Consider if the role **and terms and conditions have materially changed**?	

Further research and limitations

This chapter outlines the design of a technology-enhanced WIL package and presents a novel approach to structuring university subjects, or units, aimed at preparing future real estate professionals. A key consideration in qualitative research of this nature is the active role of the researcher, whose interpretations can influence the findings. As Levy (2006) notes, qualitative research seeks to uncover participants' perceptions and experiences, with the researcher extracting themes from this data. However, the interpretive nature of this process can lead to unintentional misrepresentation (Brown 1992).

Another limitation lies in the theoretical framing of the study, which draws on phenomenography and constructivism learning theory. These perspectives assume that knowledge is constructed through activity and social interaction, rather than transmitted directly. While widely accepted in learner-centred pedagogies, this approach may be viewed by proponents of more objectivist or pragmatic teaching methods as lacking structure or rigour (Jonassen 1991).

The chapter also introduces the AIREE framework, designed to guide the ethical adoption and integration of AI in real estate education. Although the framework is grounded in existing literature on AI in education and real estate, it has not been empirically validated. Further research is needed to assess its relevance and effectiveness in supporting educators and students in engaging with intelligent systems.

Conclusions

This chapter shared the design of an experiential learning case study, including four simulations and a supporting narrative that establishes how synchronous and asynchronous learning activities may be coordinated to provide the benefit of case-based and problem-based learning in a disrupted learning environment. When partnered with robots and drones, computer vision, LiDAR and AI, the simulations may provide a technology-enhanced work-integrated (WIL) placement for high school students with recommender systems purposed to suggest academic pathways for the student or candidate to pursue.

The related AES framework, or sequencing model, more closely aligns with education progression and has the potential to accommodate case-based and problem-based learning activities with less reliance on human resources. The model may assist universities and professional associations in bringing together their resources and materials to structure other collaborative learning programmes to advance real estate education with enabling technologies.

This chapter critically examines the role of enabling technologies, specifically those with intelligent systems, as a partner or even teacher in real estate education. It provides a means and framework for real estate educators to engage with emerging technologies and learn with AI.

Through better use and understanding of enabling technologies, we take a step towards an augmented, 'magic', learning experience that has the potential to close achievement gaps, helping non-academic students learn like their academic counterparts. That said, as our collaboration with intelligent systems deepens, it will become clearer that shared objectives and transmission of knowledge do not guarantee shared values, empathy or mutual understanding.

Questions

1 What does it take, and who is best placed, to be a future real estate professional?
2 Do we need to turn real estate education upside down, apply and break things first, before we construct the knowledge and develop the skills and competencies for the future?
3 How can real estate academics partner with intelligent persons or systems that exhibit influence and power without empathy?

Further reading

Educause Horizon Reports (https://library.educause.edu/resources/2021/2/horizon-reports)

UNESCO and national scientific research publications on AI and technology in learning and education

References

Abidoye R. B. & Chan A. P. C. (2017a). 'Valuers' receptiveness to the application of artificial intelligence in property valuation', *Pacific Rim Property Research Journal*, 23:2, 175–193. DOI: 10.1080/14445921.2017.1299453

Abidoye R. B. & Chan A. P. C. (2017b). 'Artificial neural network in property valuation: Application framework and research trend', *Property Management*, 35:5. 554–571.

AQF (2021). *The Australian Qualifications Framework (AQF)*. Retrieved October 11, 2025, from https://www.aqf.edu.au/framework/aqf-levels

Australian Property Institute (API) (2023). *Future of the property profession report*. The Australian Property Institute. Sydney NSW 2000.

Australian Property Institute (API) (2024). Rules of professional conduct, viewed 23 September 2024. https://www.api.org.au/wp-content/uploads/2024/04/Rules_of_Professional_Conduct_eff._31_March_2024.pdf

Australian Property Institute. (API) (2021). API education, website, viewed 4 March 2021. https://www.api.org.au/professional-development/

Azasu S. & Gibler K. M. (2016). The CDIO model for real estate curriculum development. *Journal of Real Estate Literature*. 24:2. 287–308. DOI: 10.1080/10835547.2016.12090429

Biggs J. (1993). From theory to practice: A cognitive systems approach. *Higher Education Research & Development*, 12:2. 73–85.

Biggs J. (1999). What the student does: Teaching for enhanced learning. *Higher Education Research & Development*. 18:1. 57–75.

Biggs J. & Tang C. (2009). *Teaching for quality learning at university: What the student does*. 3rdEd., SRHE and Open University Press. Berkshire. England.

Boyd T. (2005). Stakeholder impact on property education programs. *Proceedings from the 11th Annual Pacific Rim Real Estate Society Conference*. Melbourne. 23–27 January 2005.

Boyd S. (2015). *Playing property – designing serious games to enhance the learning experience for undergraduate property students in Australia*. Doctoral thesis. University of the Sunshine Coast. Sippy Downs. QLD.

Boyd S. (2021). Partnering with intelligent systems to enhance the practice of property valuation. *2021 Virtual ASRES-GCREC-AREUEA Joint International Real Estate Conference. Future of Real Estate: Impact of Technologies and Global Changes*. July 18–21, Singapore.

Boyd, S. (2022). The structured learning of valuation modelling, *Pacific Rim Property Research Journal*, 27, 1–18

Boyd, S. (2023). Humane Artificial Intelligence in real estate education (AIREE), PRRES Conference, January.

Boyd, S., Small, G. & Boyd, T. (2024). The residual skills to augment property valuation, PRRES Conference, January.

Braesemann F. & Baum A. (2020). PropTech: Turning real estate into a data-driven market?, viewed 18 July 2021, https://ssrn.com/abstract=3607238 or 10.2139/ssrn.3607238

Brown A. (1992). Design experiments: Theoretical and methodological challenges in creating complex interventions in classroom settings. *Journal of Learning Sciences.* 2:2. 141–178.

Charles S. (2016). Graduate real estate education: The ULI Hines student competition as a pedagogical tool. *Journal of Real Estate Practice and Education.* 19:2. 149–174.

Disability Discrimination Act 1992.

Education (Work Experience) Act 1996.

Emanuel E., Christensen G., Steinmetz A., Alcorn B., Bennett A. & Woods D. (2013). Online education: MOOCs taken by educated few. *Nature*. 503, 21 November. 342–342.

Flyability. (2024a). Amusement park inspections: Faster, safer, and more economical with the Elios, website viewed 14 November 2024, https://www.flyability.com/casestudies/amusement-park-inspections

Flyability. (2024b). Drone scan to BIM: The Elios 3 digitizes a cement plant, website viewed 14 November 2024, https://www.flyability.com/casestudies/drone-scan-to-bim

Flyability. (2024c). Infrastructure inspection drones, website viewed 14 November 2024, https://www.flyability.com/infrastructure-inspection-drones

Flyability. (2024d). Mapping Alcatraz: Digitizing the notorious prison, website viewed 14 November 2024, https://www.flyability.com/casestudies/drone-mapping-alcatraz

Government, Department of Education, Skills and Employment, Australian Qualifications Framework (AQF) (2021). AQF Levels, website viewed 12 March 2021. https://www.aqf.edu.au/aqf-levels

Institute of Electrical and Electronics Engineers. (2017). Ethically aligned design a vision for prioritizing wellbeing with artificial intelligence and autonomous systems, Version 2, report, viewed 11 September 2022, http://www.standards.ieee.org/develop/indconn/ec/autonomous_systems.html

Jimenez L. & Boser U. (2021). *Future of Testing in Education: Artificial Intelligence*, Center for American Progress, report, viewed 11 September 2022, https://www.americanprogress.org/article/future-testing-education-artificial-intelligence/#:~:text=challenges%20of%20AI-,Artificial%20intelligence%20can%20help%20students%20learn%20better%20and%20faster%20when,the%20naked%20eye%20cannot%20see

Jonassen D. (1991). Objectivism versus constructivism: Do we need a new philosophical paradigm? *Educational Technology Research & Development*. 39:3. 5–14.

Kitchin, R. (2014). *The data revolution: Big data, open data, data infrastructures and their consequences*. SAGE Publications.

Kolb D. (1984). *Experiential learning: Experience as the source of learning and development*. Prentice Hall. Englewood Cliffs, NJ.

Kumar G., Singh G., Bhatanagar V. & Jyoti K. (2019). 'Scary dark side of artificial intelligence: A perilous contrivance to mankind'. *Humanities & Social Sciences Reviews*, 7:(5). 1097–1103.

Lattimore F., O'Callaghan S., Paleologos Z., Reid A., Santow E., Sargeant H. & Thomsen A. (2020). *Using artificial intelligence to make decisions: Addressing the problem of algorithmic bias* (Technical paper). Australian Human Rights Commission. Retrieved August 15, 2025, from https://nla.gov.au/nla.obj-2973584333/view

Levy D. (2006). Qualitative methodology and grounded theory in property research. *Pacific Rim Property Research Journal*. 12:4. 369–388.

Luppicini R. & So A. (2016). A technoethical review of commercial drone use in the context of governance, ethics, and privacy. *Technology in Society*, 46, 109–119.

Maini V. & Sabri S. (2017). Machine learning for humans. Online: https://medium.com/machine-learning-for-humans

Marton F. & Saljo R. (1976a). On qualitative differences in learning: I – outcome and process. *British Journal of Educational Psychology*. 46:1. 4–11. Scottish Academic Press. Edinburgh.

Marton F. & Saljo R. (1976b). On qualitative differences in learning: II – outcome as a function of the learner's conception of the task. *British Journal of Educational Psychology*. 46:2. 4–11, Scottish Academic Press, Edinburgh.

McGrath K., Wang B., Jackson B., Kämpf-Dern A., Malone K., Funk D. & Geurts T. G. (2020). The future of real estate education: A multi-faceted perspective. *Journal of Real Estate Practice and Education*. 22:1. 40–55. DOI: 10.1080/15214842.2020.1757354

MIT media Lab and Scalable Cooperation. (Unknown). Norman World's first psychopath AI, website, viewed 30 June 2021, http://norman-ai.mit.edu/

Mooya M. M. (2015). The education and professional practice of valuers in South Africa: A critical review. *Property Management*. 33:3. 245–274. DOI: 10.1108/PM-04-2014-0021

Newell C. (2017). The use of 'drones' in marketing a property for sale. *REIQ Journal* 2017. 35–37.

Palm P. & Pauli K. S. (2018). Bridging the gap in real estate education: Higher-order learning and industry incorporation. *Journal of Real Estate Practice and Education*. 21:1. 59–75. DOI: 10.1080/10835547.2018.12091777

Parker D. (2012). Property Education in Australia: Themes and issues, PRRES Conference, January.

Păvăloaia V. D., & Necula S. C. (2023). Artificial intelligence as a disruptive technology—A systematic literature review. *Electronics* 2023:12. 1102. DOI: 10.3390/electronics12051102

Petronzi R. & Petronzi D. (2020). The Online and Campus (OaC) model as a sustainable blended approach to teaching and learning in higher education: A response to COVID-19. *Journal of Pedagogical Research.* 4:4. 498–507.

Poon J. (2014). A cross-country comparison on the use of blended learning in property education. *Property Management.* 32:2. 154–175. DOI: 10.1108/PM-04-2013-0026

Porter L, Fields D., Landau-Ward A., Rogers D., Sadowski J., Maalsen S., Kitchin R., Dawkins O., Young G. & Bates L. K. (2019). 'Planning, land and housing in the digital data revolution/the politics of digital transformations of housing/digital innovations, PropTech and housing – the view from Melbourne/digital housing and renters: Disrupting the Australian rental bond system and tenant advocacy/prospects for an intelligent planning system/what are the prospects for a politically intelligent planning system?', *Planning Theory & Practice*, 20:4. 575–603. DOI: 10.1080/14649357.2019.1651997

Porter D. B., Campbell C., Jones H., Logan-Fleming D. & Sankey M. (Eds.). (2024). *Innovation through disruption in Australasia: The 2024 Contextualising Horizon report.* ASCILITE. https://ascilite.org/get-involved/contextualising-horizon/

Sepasgozar S., Forsythe P. & Shirowzhan S. (2016). Scanners and Photography: A combined framework. *In Australian Universities Building Education Association Annual Conference*; Central Queensland University. Brisbane, Australia.

Sepasgozar S., Davis S. R. & Loosemore M. (2018). Dissemination practices of construction sites' technology vendors in technology exhibitions. *Journal of Management in Engineering*, 34. 04018038.

Sheehy B., Boddy C. & Murphy B. (2020). Corporate law and corporate psychopaths. Psychiatry. *Psychology and Law*, (June 26, 2020), https://ssrn.com/abstract=3668064

Southgate E., Blackmore K., Pieschl S., Grimes S., McGuire J. & Smithers K. (2018). *Artificial intelligence and emerging technologies (virtual, augmented and mixed reality) in schools: A research report.* University of Newcastle, Australia. Newcastle. https://www.education.gov.au/supporting-family-school-community-partnerships-learning/resources/ai-schools-report

Starr C. W., Saginor J. & Worzala E. (2020). 'The rise of PropTech: Emerging industrial technologies and their impact on real estate', *Journal of Property Investment & Finance* 39:2. 157–169. DOI: 10.1108/JPIF-08-2020-0090

Susilawati C. & Yam S. (2013). *Property case competition: Multi-stakeholder reflection.* PRRES Conference, January.

Takin M., Peng J., Sepasgozar S. & Ebrahimi H. (2017). A Framework for Using Advanced Visualization Tools for Residential Property Management. In *Proceedings of the International Symposium on Automation and Robotics in Construction*, ISARC, Taipei, Taiwan, 28 June–1 July; Vilnius Gediminas Technical University, Department of Construction Economics & Property: Vilnius, Lithuania.

The State of Queensland (Department of Education). (2022). Work experience placements for school students procedure, viewed 23 September 2024, https://ppr.qed.qld.gov.au/pp/work-experience-placements-for-school-students-procedure

Ullah F., Sepasgozar S. & Wang C. (2018). A systematic review of smart real estate technology: Drivers of, and barriers to, the use of digital disruptive technologies and online platforms. *Sustainability* 10:9. 3142. DOI: 10.3390/su10093142

Ullah F., Sepasgozar S., Shirowzhan S. & Davis S. (2021a). 'Modelling users' perception of the online real estate platforms in a digitally disruptive environment: An integrated KANO-SISQual approach', *Telematics and Informatics*, 63. 101660, ISSN 0736-5853. DOI: 10.1016/j.tele.2021.101660

Ullah F., Sepasgozar S., Thaheem M. & Al-Turjman F. (2021b). 'Barriers to the digitalisation and innovation of Australian Smart Real Estate: A managerial perspective on the technology non-adoption', *Environmental Technology & Innovation*, Vol. 22. 101527, ISSN 2352-1864. DOI: 10.1016/j.eti.2021.101527

United Nations, Department of Economic and Social Affairs (UN). (2006). Convention on the Rights of Persons with Disabilities (CRPD). *Article 1 – Purpose*, viewed 23 September 2024, https://www.un.org/development/desa/disabilities/convention-on-the-rights-of-%20persons-with-disabilities/article-1-purpose.html

Valier A. (2020). 'Who performs better? AVMs vs hedonic models', *Journal of Property Investment & Finance* 38:3. 213–225. DOI: 10.1108/JPIF-12-2019-0157

Valuers Registration Act 1992.

Valuers Registration Board of Queensland. (2019). Guidelines for registration as a valuer in Queensland, viewed 26 September 2024, https://vrbq.qld.gov.au/wp-content/uploads/2019/07/Guidelines-for-Registration-as-a-Valuer.pdf

Vos N., van der Meijden H. & Denessen E. (2011). Effects of constructing versus playing an educational game on student motivation and deep learning strategy use. *Computers & Education.* 56:1. 127–137.

Wagner D. N. (2021). 'Artificial Intelligence and the dark side of management', *ROBONOMICS: The Journal of the Automated Economy*, 1. 10. https://journal.robonomics.science/index.php/rj/article/view/10/2

Wilkinson S., Halvitigala D. & Antoniades H. (2018). Educators, professional bodies and the future of the valuation profession. *Property Management.* 32:2. 154–175. DOI: 10.1108/PM-04-2017-0027

Young K. L., Phillips C. K., Reyes M. D., O'Connor T. C., Schierman R. S., Melde A., English E. S. & West W. N. (2024). Mission results: Creating a 3D map of a very high radiation confined space using the LiDAR-equipped Elios 3 Drone, *Nuclear Science and Engineering*. DOI: 10.1080/00295639.2024.2364463

Chapter 10

The future of valuation education

Incorporating augmented reality tools in training and skill development for valuers

Maximilian Neuger and Lynne Armitage

Introduction

Property valuation forms a central element of the property industry and therefore plays a pivotal role in a wide array of urban and economic processes. From facilitating property transactions and securing mortgages to informing corporate balance sheets and tax assessments, precise and dependable property valuations are indispensable to mature property markets and, by extension, to the broader built environment ecosystem. Traditionally, this process has relied heavily on the human expertise of professional property valuers, who synthesise a professional judgement of comparative market analyses with on-site inspections and combine their knowledge of local market conditions to reconcile a reasoned opinion of value. Fundamental to the training of a valuer of real property assets is the role of valuation education in setting standards for professional valuation practice.

Like many other industries, property valuation is dynamic and not immune to the transformational effect of technological innovation. Recent decades have simultaneously given rise to 'non-standardised' valuation methods in response to growing computational power and advancements emerging from outside the traditional domain of property valuation. Hedonic price models (HPM), for instance, are used for property price indices or the creation of Automated Valuation Models (AVMs), a method widely discussed in the literature (Neuger & Susilawati, 2025).

The concept of market value has been articulated in *Spencer v Commonwealth* (1907) and further reinforced by professional guidelines such as the International Valuation Standards (IVS), the Royal Institute of Chartered Surveyors (RICS), and the Australian Property Institute (API). Although the foundational principles of valuation remain anchored in longstanding legal precedents, the tools and techniques employed by valuers have continuously evolved throughout time. The integration of technologies into valuation education extends beyond the adoption of new tools and signifies a paradigm shift in how valuation knowledge is conveyed and acquired. Critically employed, these technologies can complement traditional teaching and learning methods

DOI: 10.1201/9781003539254-13

by offering dynamic, experiential learning environments tailored to the complexities of modern valuation practice.

Among these, Augmented Reality (AR) stands out for its capacity to overlay digital information onto physical spaces. The Vision Pro[1] released in early 2024 by Apple, is one of the most promising innovations in the realm of augmented reality. For valuation-related practices, this offers a new realm of valuation education, translating into greater opportunities for experiential learning, allowing students to engage in scenario-based exercises that replicate real-world challenges. Robust evidence supports that real-world experience and immersive learning environments lead to improved training outcomes, enhanced skill acquisition, and greater learner engagement (Cowling & Birt, 2018). If thoughtfully embedded within pedagogical frameworks, technologies akin to the Apple Vision Pro can at least play an assistive role in supporting immersive and interactive learning experiences, complementing traditional education.

When applied to valuation education, AR offers immersive and interactive learning experiences that replicate real-world tasks such as property inspections, due diligence, and market analysis. Used thoughtfully, AR can support traditional classroom teaching by creating simulation-based scenarios that help students develop spatial awareness, analytical skills, and professional judgement. For instance, students can use AR to simulate a site inspection of a commercial property, identifying visible defects like water ingress or illegal structural additions. Through digital overlays, they can assess zoning classifications in real time, evaluate lease structures by virtually exploring tenancy layouts, or compare sales data mapped onto surrounding parcels. These scenarios mirror real-world workflows and help students practise applying evidence to form a reasoned opinion of value. Importantly, such simulations can be conducted in a low-risk, repeatable environment that supports formative feedback and reflective learning. However, these technologies also introduce challenges around transparency and professional accountability, particularly if decision-making processes become overly opaque or automated. To address this, automated processes should be implemented, ensuring that the valuer's expert judgement remains central to practice. Consequently, the application of these tools must be critically assessed to ensure they support, rather than obscure, professional judgement and ethical standards.

This chapter examines how AR tools like Apple Vision Pro and mobile-based platforms can be used in the classroom. It outlines the pedagogical value of AR, links it to industry expectations, and provides examples of how institutions can integrate these technologies into valuation curricula. In doing so, the chapter contributes to a growing conversation about modernising professional education without losing sight of core valuation principles and offers new avenues for students on how these principles can be taught, experienced and internalised in an educational context. The chapter contributes to the ongoing discourse on the future of valuation education by offering both a conceptual framework and actionable recommendations for integrating AR technologies into pedagogical practice. It argues that while advanced platforms such as the Apple Vision Pro offer significant promise, their success in educational settings hinges not merely on technological sophistication but on the pedagogical readiness of educators.

Therefore, this chapter begins by critically examining the current landscape of property valuation education, commencing with a review of current pedagogical practices and their limitations. It also highlights the increasing need for technology adaptation in learning that responds to the complexity of contemporary valuation practice. The chapter also outlines key applications of AR within valuation education, including simulation-based training modules and AR-enabled assessment tools. In this context, the Apple Vision Pro is introduced as a case study example to explore the educational potential of AR in training and upskilling future valuation professionals. It also addresses essential implementation considerations, including but not limited to hardware and software requirements, content development strategies and faculty training needs. Potential challenges, for example, cost barriers to institutional inertia, are discussed.

In the Future Integration and Consideration section, the discussion extends to how AR aligns with broader technological and societal transformations, particularly within the context of the 21st century's Fifth Industrial Revolution. The chapter situates AR within the PropTech 3.0 paradigm and explores its potential convergence with the Metaverse, offering a conceptual bridge between valuation education and emerging digital ecosystems. To stimulate academic and professional discourse, this chapter includes discussion questions to prompt reflection on how valuation professionals can stay innovative by adopting new tools and anticipating industry trends.

Pedagogical gaps and the current state of valuation education

Property valuation in tertiary education has traditionally relied on a combination of classroom-based instruction and case studies. While these methods are effective for imparting foundational knowledge, these approaches often fail to provide students with practical experience to develop the critical thinking, problem-solving and analytical skills required in professional valuation judgement. In recent years, growing concerns in valuation education arose over the disconnect between academic theory and professional practice. According to the API's report on the future of valuation, 85% of respondents advocated for more practical experience within university curricula, including structured workplace placements and mentoring pathways (API, 2023).

The global COVID-19 pandemic has opened Pandora's box and further highlighted the limitations of conventional classroom-based instruction and learning and the need for more flexible, technology-driven educational solutions. The shift to remote learning sped up digital adoption in valuation education but has also exposed the need for more engaging and interactive learning tools in virtual environments. As the valuation profession navigates an increasingly uncertain and complex future, it is arguably imperative for educational institutions to equip students with the skills, flexibility and resilience required to thrive in a rapidly changing environment.

Despite growing awareness, major hurdles persist. Classroom settings struggle to replicate the complexity of real-world valuation tasks, compounded by a lag between traditional teaching methods and the pace of technological progress, leaving students unresponsive to evolving demands of the industry. The valuation industry now sits at the crossroads, yet there remains little structured guidance on how to integrate some of these digital tools into valuation education. As valuation adapts to new technologies, educators must update how foundational skills are taught. Professionals widely recognise that technology adoption offers prospective gains across multiple dimensions, from operational efficiency and valuation accuracy to financial performance (Crosby, 2000).

Traditionally, there is widespread acknowledgement that thorough valuations require human judgement, particularly in areas such as local market knowledge and a thorough inspection process; this shall not be translated into resistance to technological advancements. Scholarship suggests that emerging technologies can serve as complementary tools that support, rather than compete with, professional judgement in valuation. On-site inspections of physical properties remain essential for valuers to assess the quality of buildings, detect flaws and consider environmental factors. These traditional skills can be augmented through technology by improving efficiency, accuracy and analytical capabilities. A well-rounded strategy positions the valuation profession well for future evolution, where success is likely to depend on effectively combining professional knowledge with technological advancement. This suggests that while due diligence processes remain important, the future of valuation education must strike a better balance between academic learning and practical experiences. *Without action, the profession risks a widening skills gap and inadequate preparation of future graduates.* Educational institutions must equip students with technological literacy and practical skills needed to succeed in the modern valuation landscape.

Technological overview: AR, XR and Vision Pro

Overview of augmented reality (AR) technology

Augmented reality (AR) is a technology that overlays digital information on a real-world environment. Mixed Reality (MR) blends digital and physical elements in real time. Unlike Virtual Reality (VR), with fully immersive digital environments, AR creates a composite view that extends a person's perception and interaction with reality by overlaying virtual content onto real-world settings. VR, AR and MR fall under the term 'Extended Reality' (XR).

On 5 June 2023, at Apple's Worldwide Developers Conference, the company unveiled Apple Vision Pro, representing a significant innovation within this space, introducing high-resolution displays, spatial mapping and gesture recognition. *These features enabled more engaging, hands-free interaction with digital content. This progression marks a shift from smartphone-based AR to professional-grade headsets designed for complex tasks and learning environments.*

AR systems rely on a combination of hardware and software components with a display device typically a mobile phone, tablet or headset, tracking system, and a processing unit. LiDAR (Light Detection and Ranging) sensors are a common tracking technology used in AR headsets. LiDAR uses lasers to measure distances by timing how long reflected light takes to return. The data is used to create point clouds that digitally represent 3D objects.

Although the Apple Vision Pro is a remarkable device, is not the only viable AR solution. Table 10.1 presents a comparative overview of AR devices, from mobile-based applications to headsets. Although *devices like* Microsoft HoloLens[2] and Magic Leap represented significant progress over recent decades, this chapter references the Vision Pro as a case study due to its advanced capabilities, but the pedagogical principles discussed here are applicable across multiple AR platforms.

The introduction of ARKit by Apple and ARCore by Google in 2017 marked a significant milestone in the democratisation of AR technology. These Software Development Kits (SDKs) enable developers to easily create AR

Table 10.1 Comparative overview of selected AR devices for valuation education (Authors, 2025)

Feature	*Apple Vision Pro*	*Microsoft HoloLens 2*	*Magic Leap 2*	*Mobile AR Applications*
Display Type	Dual micro-OLED	Waveguide	Waveguide	LCD/OLED
Spatial Mapping	LiDAR + SLAM	SLAM + depth sensors	SLAM + depth sensors	Limited (via camera/ ARKit)
Gesture/Eye Tracking	Yes (Eye, Hand)	Yes (Hand)	Yes (Hand, Eye optional)	No
Field of View (FoV)	~100°	~52°	~70°	Device dependent
Integration with Curriculum	High (via Apple SDKs)	Medium (via Unity, MRTK)	Medium (Unity SDKs)	High (entry-level apps)
Suitability for Group Use	Low (1 user per device)	Medium	Medium	High (shared viewing)
Estimated Cost (AUD)	$6,000–$7,000	$5,000–$6,000	$4,500–$6,000	$1,500–$2,500

applications and deploy them on a wide range of compatible devices. The subsequent years witnessed a surge in AR adoption across various industries, with notable examples including Pokémon GO in gaming and IKEA Place in retail.

Extended reality in education and training

The value of augmented reality (AR) in higher education goes beyond its technical affordances. Educational learning theories suggest that knowledge is formed through direct interaction with the environment. From this perspective, immersive technologies are attractive for their ability to support experiential learning. In practice-oriented settings, students do not passively consume information; they explore scenarios, test assumptions and develop understanding through hands-on, contextual experiences. Cochrane et al. (2022) argue that transformative mobile learning strategies offer a particularly effective framework for integrating extended reality (XR) into professional education, aiming to foster learner autonomy, critical thinking and applied competence.

Constructivist theories emphasise active, situated engagement. Kolb's Experiential Learning Cycle, though developed in the 1980s, remains highly relevant today. It provides a structured model for designing interactive environments by progressing through concrete experience, reflective observation, abstract conceptualisation and active experimentation. This process not only deepens conceptual understanding but also reinforces practical skills. Likewise, Bloom's Revised Taxonomy offers a useful framework for aligning XR activities with cognitive development. Immersive tools such as AR can help students move beyond rote learning by enabling them to apply knowledge in realistic contexts, analyse complex information, evaluate competing approaches and generate creative solutions within digital environments.

Recent empirical research supports this pedagogical alignment. Immersive AR/VR has demonstrated promise in high-stakes fields such as healthcare and construction education, where spatial reasoning and real-world problem-solving are central. Birt & Vasilevski (2021) found that multiuser immersive VR environments significantly improved student usability scores and learning engagement compared to single-user settings. These findings indicate that collaborative, embodied experiences can simulate workplace environments and offer scalable solutions for professional training, including property valuation.

Furthermore, Doggett et al. (2021) used VR to assess the cognitive effects of acoustic design in learning environments. The study demonstrated how XR platforms could serve as controlled environments for testing spatial awareness, perception and task performance. These represent critical skills in valuation, where environmental factors can affect an opinion of market value.

Importantly, technological interventions in education must be underpinned by robust pedagogical frameworks. Cowling and Birt (2018) advocate a 'pedagogy before technology' approach, emphasising the value of Design-Based

Research (DBR) in aligning XR applications with authentic, discipline-specific learning outcomes. In their study on paramedic training, augmented reality and 3D-printed tools were combined to simulate airway management procedures, providing learners with repeatable, low-risk training environments that mirror the complexity of real-world practice. This model, focused on fostering practical competence and enabling measurable feedback, has clear relevance for valuation education, where applied judgement and procedural rigour are essential.

Educational theory suggests that students learn best by doing. Extended reality (XR) technologies offer practical and pedagogically valuable applications for professional development in property valuation, with augmented reality (AR) standing out as particularly relevant.

Valuation applications of AR tools

The core strength of AR lies in its ability to superimpose digital information onto the physical world. AR allows students to explore digital property data in a real-world context. This helps them practise skills such as identifying property features, assessing external influences and interpreting site-specific information. Unlike Virtual Reality (VR), which immerses users in entirely virtual environments, AR maintains a vital connection to actual physical spaces. This capability makes AR especially well-suited to valuation education, where accuracy in interpreting physical attributes and situating the property within its location-specific factors and built environmental context is critical.

Due to the high cost and logistical barriers of using AR headsets like Vision Pro in large classes, institutions could rotate device use among small student groups. Others can engage in theory or analysis during these sessions. Educators may also conduct instructor-led demonstrations with single devices mirrored to classroom displays to introduce core valuation concepts. Alternatively, entry-level AR platforms, such as ARKit-enabled iPads or smartphones, can offer simplified but effective AR experiences. These mobile solutions lower the financial barrier to entry and can serve as introductory tools before advancing to headset-based simulations.

AR simulations for training new valuers

Site inspections are a cornerstone of real property valuation practice, where the date of inspection represents the date of valuation. However, educational opportunities for field-based learning are frequently constrained by logistical challenges and resource limitations.

Students can virtually explore buildings, systematically assessing construction quality, identifying structural defects and interpreting site-specific features. Real-time digital prompts guide learners by highlighting important elements, such as signs of structural fatigue, illegal modifications, non-compliant structures or maintenance issues like efflorescence or water ingress. For retail assets,

AR can help cadet valuers visualise internal layouts, assess lease configurations and evaluate tenant visibility.

These simulations align closely with Kolb's Experiential Learning Cycle, particularly addressing stages of concrete experience and active experimentation. By interacting with digital overlays, students bridge theoretical concepts with practical application. These experiences strengthen their observational skills, spatial accuracy and diagnostic reasoning. Such immersive experiences enable students to learn through trial and error in a low-risk environment, providing immediate, formative feedback that fosters professional judgement and analytical thinking.

While AR simulations greatly strengthen educational accessibility and consistency, they complement rather than replace the necessity of field-based experience. Real-life valuation exercises remain essential, but AR provides foundational training that prepares students effectively for supervised professional practice. Furthermore, AR simulations can replicate complex market situations and varied property types, offering students exposure to scenarios they might otherwise rarely encounter during limited university placements or cadetships.

The capacity to digitally overlay information while maintaining a connection to the physical environment makes AR a promising tool for training new valuers. AR simulations directly support the core competencies defined by the API Rules of Professional Conduct (2021), particularly in systematic property inspection and due diligence practices. By enabling students to perform simulated assessments aligned with professional workflows, these educational experiences strengthen students' abilities to conduct thorough and compliance and regulatory standards in property valuations.

Immersive learning experiences for exploring complex valuation scenarios

Accurate property valuation requires a comprehensive understanding of market dynamics, including zoning regulations, infrastructure developments, demographic shifts and recent comparable sales. Traditional educational settings often struggle to provide applied experiences that help students effectively interpret and integrate these contextual factors.

The capabilities of AR can facilitate the process by overlaying real-time digital representations of zoning changes, infrastructure upgrades and recent comparable sales data onto actual properties or simulated environments. Students equipped with AR headsets can visually interact with market information *in situ*, enabling them to evaluate the impact of external factors directly on property values. For instance, learners can observe and assess how zoning restrictions or planned infrastructure improvements might influence the valuation outcomes for specific properties. Additionally, AR's multiuser capabilities allow students and instructors to collaboratively explore these scenarios, enhancing learning through real-time discussion and joint analysis.

This immersive AR environment significantly strengthens learners' ability to interpret and synthesise complex market data, aligning directly with the higher-order cognitive skills outlined in Bloom's taxonomy, specifically, the stages of analysis and evaluation. The active engagement with contextual overlays and collaboratively resolving valuation challenges facilitates students' development of critical thinking and effective communication skills. This approach reinforces professional judgement, situational awareness and methodological rigour consistent with the API's guidelines for evidence-based market analysis and value reconciliation.

While AR provides powerful tools for visualising market contexts, it must serve as an educational aid rather than a replacement for professional judgement. The responsibility for evidence selection, analytical interpretation and valuation decisions remains firmly with the valuer. In educational simulations, instructors can observe and provide real-time feedback on students' analytical processes, further reinforcing correct valuation methodologies and professional standards. However, it is crucial to clarify that actual professional valuation practice necessitates direct property inspection and the application of personal expertise, in line with the API Rules of Professional Conduct and the International Valuation Standards.

Effective use of Vision Pro in market analysis training involves structured scenarios that mimic real-world complexities. Educators can design specific simulations where students interact with spatially contextualised data, practice evidence ranking and justify their analytical choices. Additionally, the capability to record and review AR sessions facilitates thorough performance assessments, aiding in continuous improvement and standardisation of best practices across educational programmes. Its unique combination of real-world context and digital augmentation positions AR as an ideal medium for training the next generation of valuation professionals.

AR-based assessment tools for evaluating skills and knowledge

Beyond immersive learning, AR technologies also offer new approaches to assessment. These tools support not only how students learn but also the evaluation of applied skills in realistic scenarios. As the industry contends with growing complexity driven by evolving sustainability requirements, technological change and shifting market dynamics, educational approaches must prepare future valuers to navigate these challenges with both technical proficiency and contextual awareness.

Effective valuation education requires robust assessment methods capable of evaluating professional judgement and applied competencies under realistic conditions. Traditional written exams and static case studies typically fall short in capturing the dynamic, multifaceted nature of valuation tasks encountered in practice.

Apple Vision Pro provides advanced assessment tools that leverage AR capabilities to create interactive, scenario-based evaluations. In these assessments, students engage with virtual property inspections, evidence evaluation and valuation reasoning tasks that closely replicate real-world conditions. The AR system meticulously tracks user interactions, including visual attention patterns, evidence selection processes and the rationale behind valuation decisions. For instance, during a simulated property inspection, the system records the student's inspection path, identifies focal points and notes the assessment of critical property features such as structural defects, external influences or zoning restrictions. Additionally, virtual portfolio scenarios allow students to select and justify comparable sales, with the AR system tracking their evidence prioritisation and analytical logic.

AR-based assessments offer authentic performance evaluation aligned with practical competencies outlined by the API and IVS standards. By capturing detailed interaction metrics, educators can provide targeted, formative feedback on students' observational accuracy, analytical precision and professional reasoning. These evaluations support higher-order cognitive processes, corresponding to Bloom's taxonomy stages of analysis, evaluation and creation, where students synthesise complex data, articulate valuation rationales and defend their professional conclusions.

Educational institutions can integrate AR assessment scenarios into curricula by designing structured simulations that progressively test students' valuation skills across various contexts and complexity levels. Continuous formative assessments facilitated by AR tracking ensure consistent application of valuation methodologies and support the professional development of critical judgement skills required by industry standards. While AR supports immediate educational needs, it also prompts broader reflections about the future of valuation education and the technologies reshaping the profession.

Future integration and considerations

PropTech, the fourth industrial revolution and the future of valuation education

PropTech, a *portmanteau* for property technology, represents the digital transformation of the property industry through innovative technologies. PropTech 3.0 is emerging as 'the future of real estate', characterised by advanced technologies that promise to reshape the industry. This evolution has occurred in three waves and introduced new forms of how real estate professionals operate. While much of this innovation has occurred in the investment, development and asset management sectors, there has also been relevance to valuation practices. Table 10.2 provides an overview of these waves, with core technologies and impact on the real estate sector.

Table 10.2 Evolution of PropTech from 1.0 to PropTech 3.0 (Authors, 2025)

PropTech Wave	*Period*	*Core Technologies*	*Key Impact for the Real Estate Sector*	*Relevance to Valuation Education*
1.0	1980–2000	Rise in computing power, Advent of spreadsheets, early database systems (Excel, Argus, Yardi)	Digitised valuation, investment and property management processes	Introduced foundational modelling and data analysis tools still taught in core curricula
2.0	2008–2019	Cloud computing, mobile wave through apps, sensors and the rise of the sharing economy (Airbnb, WeWork, Uber)	Big Data, Digital Real Estate and Smart Buildings	Provided richer market context and new data inputs for analysis and training
3.0	2020–present	AI, blockchain, XR/AR/VR, Metaverse, robotics	"Future of Real Estate"	Opens the door for experiential, simulation-based learning in valuation education

While real estate has traditionally lagged in digital adoption, new technologies are prompting a shift. The property sector has been one of the last industries to adopt technological change, despite being the largest asset class globally (Baum, 2017). 'Real estate' describes land and the improvements to the land, whereas 'real property' describes a bundle of rights and interests associated with property. Real estate, traditionally a 'brick and mortar' asset, is being jeopardised by the rise of digital assets and virtual realities. This new 'bits and bricks' frontier, in which real property and virtual real estate converge, challenges how value is represented and transacted across both physical and digital domains. Digital assets are not currently recognised as legal interests in real property. As such, they are excluded from valuation under standards like the International Valuation Standards (IVS), the Royal Institute of Chartered Surveyors (RICS) or the Australian Property Institute (API). As such, the question emerges: how should valuation educators approach emerging technologies like AR and the Metaverse without compromising methodological rigour?

As the valuation industry continues to evolve in response to technological advancements and changing market demand, the role of AR in education is likely to become increasingly critical. By embracing these technologies and their transformative potential, the valuation community can position itself at the forefront of educational innovation, ensuring a well-prepared and adaptable workforce to meet future challenges. However, realising its full potential requires ongoing collaboration and knowledge sharing among educators, researchers, industry partners and policymakers.

How does the valuation industry adapt to PropTech 3.0?

Technological innovation increasingly disrupts property practices. The JLL Future Vision report (2024) predicts that through 2030 and beyond, the human-machine interactions of the 5th Industrial Revolution will alter how and where humans work, live, play and invest, with significant ramifications for real estate stakeholders. Blockchain technology has enabled new forms of property transactions and ownership records. Artificial intelligence is an enhanced property analysis and decision-making process. Virtual, augmented and mixed reality technologies, along with the Metaverse, are transforming the visualisation and experience of properties. The integration of 3D printing and robotics is revolutionising construction processes, whereas quantum computing promises unprecedented computational capabilities. Table 10.3 presents some of these key technologies with their relevance to valuation.

How does AR integrate with other systems, changing the built environment[3]

The Metaverse, or Web 3.0, as the next generation of the internet, has emerged as a concept describing persistent, immersive digital environments, often supported by VR, AR and spatial computing. Spatial computing influences how

Table 10.3 Disruptive technologies in the built environment (Authors, 2025)

Technology	*Core Function*	*Relevance to Valuation*
AI	Pattern recognition, automation, large-scale analysis	Supports valuation modelling and risk profiling
AR	Overlay of digital data on physical environments	Supports inspections, communication and spatial analysis
IoT (OECD, 2016:80) (Duarte & Alvarez, 2019)	Sensor-based data monitoring	Enriches property intelligence (e.g., energy, occupancy)
3D Printing	Additive construction and modelling	Impacts future highest and best use (HBU) and cost-based valuation scenarios
Quantum Computing, AVs, Smart Cities	High-complexity computation and urban infrastructure	Speculative

humans interact with the built environment, offering new ways of engaging with virtual spaces. Technologies like AR and VR are changing how users interact with digital representations of real-world spaces. In education, this means students can now explore virtual models of buildings or neighbourhoods as part of their learning or devices like AR headsets let users view and interact with digital property data in real space.

According to Andrew Phipps, Head of Business Development at Cushman & Wakefield EMEA, the Metaverse presents enriching opportunities for customer experience and workplace innovation, particularly for organisations that utilise office space (Cushman & Wakefield, 2022). *JLL's "Metamall"* enables businesses to lease or purchase virtual spaces for retail operations (JLL, 2022). *CBRE anticipates growing demand for digital architecture and virtual portfolio management* (CBRE, 2022).

This represents an opportunity for real estate professionals to develop new competencies that support both the physical and virtual environments. However, industry experts maintain a balanced perspective of the Metaverse's role. Abhishek Bajpai, Managing Director of Occupier Services at Colliers Asia, notes that while the Metaverse will not replace physical assets, it will increasingly create synergies with the real world (Colliers, 2022).

VR and AR support the creation of digital twins as immersive replicas of real-world assets. Digital twins are detailed 3D models of real properties, used to simulate inspections or design changes, or AR tools can be used to display building plans or simulate inspections based on real-world models. This raises important pedagogical questions about the boundaries between digital representations and the legal standards underpinning real property valuation.

These developments offer opportunities to cultivate digital literacy and critical awareness among valuation students, particularly in relation to the ethical use of emerging technologies. In educational contexts, exposure to the Metaverse may help students build adaptability and awareness of emerging digital property environments in a rapidly evolving industry landscape. The promise of the Metaverse lies in its capacity to supplement, but not supplant, the foundational principles of valuation.

Metaverse and real estate: Quo vadis? Promise or peril[4]

For valuation education, the challenges presented are dual. Students must be prepared for a future where digital tools shape industry practices, client expectations and communication environments, while simultaneously reinforcing the importance of professional judgement, evidence-based reasoning and compliance with regulatory frameworks. Regulation lags innovation. Immersive tools can improve teaching, but the process of estimating market value still relies on professional judgement and standards. As such, as of now, the Metaverse is best treated as a topic for contextual awareness, offering insight into future service demands and the digital property ecosystem, rather than a methodological innovation in valuation itself.

Despite the benefits of AR in valuation education, its widespread adoption is constrained by cost, access to hardware and faculty preparedness. Advanced headsets like the Apple Vision Pro, with costs ranging from AUD $6,000 to $7,000 per unit, are unlikely to be feasible for most institutions at scale. In a class of 30 students, for example, even five headsets would require an upfront investment exceeding AUD $35,000. How can educational institutions ensure equitable access to immersive technologies for students across varying socio-economic contexts?

Immersive AR technologies offer educators the opportunity to design learning experiences that reflect professional realities. These tools can deepen applied skills, improve engagement and better prepare students for the evolving demands of valuation practice.

Effective implementation requires instructors to design scenario-based learning outcomes and professional standards. To support this, institutions should invest in faculty training initiatives such as professional development workshops, peer mentoring programmes and access to trial devices. Collaborative curriculum design and hands-on exposure to AR tools will help educators build confidence and competence in deploying immersive technology effectively. Without this foundation, the transformative potential of AR in valuation education may not be fully realised. Taken together, these developments suggest a clear trajectory: the valuation profession must embrace educational innovation that remains grounded in core principles while adapting to a digital future.

Conclusion

The integration of augmented reality (AR) into valuation education presents an opportunity to modernise traditional pedagogical approaches and align learning experiences with the demands of an increasingly digitalised property sector. This chapter suggests that these tools improve student engagement, skill acquisition and knowledge retention by enabling active and experiential learning that directly supports established educational frameworks.

Based on this premise, AR implementation in valuation education moves beyond technological novelty to create meaningful learning experiences that develop both technical competencies and critical thinking capabilities, which are both essential for contemporary property practice. These technologies strengthen how core concepts are taught, but professional judgement, critical thinking and ethical responsibility, guided by standards such as the API Rules of Conduct and International Valuation Standards that define valuation, remain paramount to the valuation process.

Nevertheless, several constraints must be acknowledged. *Successful implementation depends on structured instructional design, faculty development, and institutional readiness. AR should be introduced incrementally, with attention to device cost, access equity, and learning outcomes.* High implementation costs, limited faculty readiness in staff education, awareness and equipment and varied user acceptance may hinder adoption, particularly in resource-constrained settings. As such, AR should be implemented with careful attention to instructional design and learning outcomes. It must serve as a supplement to, not a substitute for, traditional instruction and the mentorship of experienced professionals. The adoption of AR in valuation education is not a *panacea* but rather a means that must be thoughtfully integrated with existing pedagogical practices and aligned with the evolving needs of the industry. Educators must balance foundational pedagogy with new technological opportunities. With careful integration, AR can become a powerful supplement to traditional methods, supporting both skill development and academic standards.

Although certain challenges may exist in implementation, the potential benefits described in this chapter in terms of enhanced skill acquisition, knowledge retention and practical experience make this a promising field for further exploration and development. *Moving forward, collaboration among educators, industry leaders, and policymakers will be key. Shared best practices, trials, and applied research will help shape scalable, standards-aligned approaches to valuation education. In embracing these tools responsibly, the valuation profession can ensure its next generation of practitioners are both technically proficient and ethically grounded.*

Summing up the chapter, proposed research questions are offered to guide future inquiry and scholarly discourse.

1 *To what extent does the integration of AR-based simulations improve student performance in core valuation competencies compared to traditional instructional methods?*
2 *What are the perceived barriers among educators to integrating AR into professional valuation curricula?*
3 *How do immersive technologies affect student confidence, decision-making quality, and retention of valuation-specific knowledge?*

This chapter contributes to the growing discourse on the future of valuation education but also the broader profession by offering both conceptual grounding and practical insights into AR's potential. While this chapter presents a conceptual foundation, further empirical research is required to evaluate the efficacy, scalability and pedagogical value of AR technologies. Ultimately, the success of AR in valuation education hinges on a collaborative effort among educators, researchers, industry practitioners and policymakers. In embracing innovation while preserving professional standards, the valuation community can ensure a more adaptive, skilled and future-ready workforce, well-equipped to meet the challenges of the evolving built environment.

Notes

1 This chapter uses the term Augmented Reality (AR) to refer to technologies that overlay digital content onto the physical environment. While Apple markets its Vision Pro headset as a 'spatial computing device,' in this chapter, it is discussed as a high-end example of AR.
2 The Successor the HoloLens 2 was released in 2019.
3 Get set for the 5th Industrial Revolution (JLL, 2024).
Five Megatrends (PWC, 2016).
The Great Instability: Mega Trends for the 21st Century(BNP Paribas, 2021).
3 Megatrends Shaping The Future Of Global Real Estate (CBRE, 2018).
4 Recommendations: Value creation in the metaverse The real business of the virtual world (McKinsey & Company, 2022), Building The Metaverse (Jon Radoff, 2021).
The Next Frontier: CRE and the Metaverse (Cushman & Wakefield, 2022).
Retail, metaverse and metamall (JLL, 2022).
Enter the Metaverse (CBRE, 2022).
Expert Talks | Real Estate In The Metaverse (Colliers, 2022).

References

API. (2023). The Future of the Australian Property Valuation Industry. Australian Property Institute. Retrieved October 11, 2025, from https://www.api.org.au/professional-development/knowledge-hub/future-of-the-property-profession-report/

Baum, A. (2017). PropTech 3.0: The Future of Real Estate. Saïd Business School, University of Oxford. Retrieved October 11, 2025, from https://www.sbs.ox.ac.uk/sites/default/files/2018-07/PropTech3.0.pdf

Birt, J. R., & Vasilevski, N. (2021). Comparison of Single and Multiuser Immersive Mobile Virtual Reality Usability in Construction Education. *Educational Technology & Society* (Vol. 24, Issue 2, pp. 93–106). https://psycnet.apa.org/record/2021-40231-007

BNP Paribas. (2021). The Great Instability: Mega Trends for the 21st Century. Retrieved October 11, 2025, from https://viewpoint.bnpparibas-am.com/the-great-instability-mega-trends-for-the-21st-century/

CBRE. (2018, March 15). *3 Megatrends Shaping The Future of Global Real Estate*. Forbes. https://www.forbes.com/sites/cbre/2016/07/05/3-megatrends-shaping-the-future-of-global-real-estate/?nowelcome=&sr_source=lift_facebook&utm_campaign=cbre&utm_medium=referral&utm_source=facebook&sh=22a814951ea1

CBRE. (2022). Enter the Metaverse: The Rise of Digital Worlds and Their Implications on Commercial Real Estate. Retrieved October 11, 2025, from https://www.cbre.com.sg/insights/articles/enter-the-metaverse

Cochrane, T., Narayan, V., Aiello, S., Alizadeh, M., Birt, J., Bone, E., Cowie, N., Cquniversity, M. C., Deneen, C., Goldacre, P., Sinfield, D., Stretton, T., & Worthington, T. (2022). Analysing Mobile Learning Designs: A Framework for Transforming Learning Post-COVID. In *Australasian Journal of Educational Technology* (Vol. 38, Issue 4, pp. 1–21).

Colliers. (2022). Real Estate in the Metaverse: Augmenting Reality: The Metaverse Opens a World of Real Estate Without Limits. Retrieved October 11, 2025, from https://www.colliers.com/en-xa/news/e22-expert-talks-real-estate-in-the-metaverse

Cowling, M., & Birt, J. (2018). Pedagogy before Technology: A Design-based Research Approach to Enhancing Skills Development in Paramedic Science Using Mixed Reality. *Information (Switzerland)*, *9*(2), 29. https://doi.org/10.3390/info9020029

Crosby, N. (2000). Valuation Accuracy, Variation and Bias in the Context of Standards and Expectations. *Journal of Property Investment & Finance*, *18*, 130–161. https://doi.org/10.1108/14635780010324240

Cushman & Wakefield. (2022). The Next Frontier: CRE and the Metaverse. Retrieved October 11, 2025, from https://www.cushmanwakefield.com/en/insights/the-next-frontier-commercial-real-estate-and-the-metaverse

Doggett, R., Sander, E. J., Birt, J., Ottley, M., & Baumann, O. (2021). Using Virtual Reality to Evaluate the Impact of Room Acoustics on Cognitive Performance and Well-Being. *Frontiers in Virtual Reality*, *2*. https://doi.org/10.3389/frvir.2021.620503

Duarte, F., & Alvarez, R. (2019). The Data Politics of the Urban Age. *Palgrave Communications*, *5*, 1–7. https://api.semanticscholar.org/CorpusID:167220784

JLL. (2022). Will We Ever Really Work in the Metaverse? Retrieved October 11, 2025, from https://www.jll.com/en-au/insights/will-we-ever-really-work-in-the-metaverse

JLL. (2024). The Future of Real Estate: New Perspectives on a Fast-changing World. Retrieved October 11, 2025, from https://www.jll.com/en-au/insights/market-outlook/future-vision-real-estate

Jon, Radoff. (2021, February 25). Building the Metaverse. https://Medium.Com/Building-the-Metaverse

McKinsey & Company. (2022). Value Creation in the Metaverse. Retrieved October 11, 2025, from https://www.mckinsey.com/capabilities/growth-marketing-and-sales/our-insights/value-creation-in-the-metaverse

Neuger, M., & Susilawati, C. (2025). Urban Megaprojects from Isolation to Integration: A Property Market Perspective on Flagship Buildings. *Buildings*, *15*(7), 1156. https://doi.org/10.3390/buildings15071156

OECD. (2016). The Internet of Things: Seizing the Benefits and Addressing the Challenges. Retrieved October 11, 2025, from https://www.oecd.org/en/publications/the-internet-of-things_5jlwvzz8td0n-en.html

PWC. (2016). Five Megatrends and their Implications for Global Defense & Security. Retrieved October 11, 2025, from https://www.pwc.com/gr/en/publications/government/five-megatrends-implications-global-defense-and-security.html

Section 4

Reflections on legal, ethical and future directions in digital valuation

Chapter 11

A(I) legal perspective

Process, procedure, risk and GenAI use

Lucy Cradduck

Introduction

Generative artificial intelligence tools (GenAI) are but the current iteration of the technological changes that began centuries ago. Comparatively, a core impact of the Industrial Revolution was that many tasks previously undertaken by individuals are now undertaken by machines. In many instances, the role of the individual effectively changed from that of 'active doer' to the supervisor of the machine, which also led (ultimately) to a reduction in the number of individuals employed in a particular industry. This is seen, for example, in the mechanisation of weaving and in the slow decline of handloom weavers (Nardinelli, 1986). There remains, however, a need for the active engagement of individuals in the weaving process: in designing patterns, in ensuring the machines do not stop working, in loading the materials, in setting up the looms for a particular pattern, and in checking the final product (Kyosev and Boussu, 2022).

In an office context, various tasks, such as typing reports, were made less manual through the typist's use of electronic computers, which enable the easier correction of, and additions to, documents (Pigozzi, 2023). More recent technological advances mean that computers can be both 'co-editors' and content creators. This is seen, for example, in the 'autocorrect' function of word processing applications, which, unless you switch them off, continually seek to 'assist' by providing suggestions for finishing the word or phrase the user is typing and in the writing assistant functionalities designed to assist the user to check their spelling and grammar (Dale and Viethen, 2021). However, despite the widespread use of these types of tools, the role of the individual as both author and editor has not (yet) been completely displaced (Baron, 2024).

In a data analysis and computational context, the human relationship with computers has come full circle. The original 'computers' that undertook the significant mathematical calculations that enabled the early groundbreaking voyages of the NASA astronauts were not machines but incredibly intelligent and skilled women (Pavlić, 2024). That role is now reversed, as it is individuals as users who engage electronic computers and software to perform such calculations and other functions (Baecker, 2014). Similarly, the 'thinking machine'

DOI: 10.1201/9781003539254-15

as a concept is not new (see discussion of Turing, 1950, as considered in Murugesan, 2025). The increased development of the capacities of AI tools enables computers and devices now to 'think', that is to analyse and create data. It is this capacity that leads to concerns specifically as it relates to the accuracy and thus the reliability of GenAI-produced outputs.

The ability of GenAI to create data, rather than merely providing faster analysis of materials and data already gathered, must be balanced against several risks. Absent critical engagement with any GenAI output, any reliance on the outputs of GenAI raises several legal concerns, which can place a valuer at risk of both disciplinary and negligence actions. This chapter will provide an overview, from a legal perspective, of the issues facing the valuation profession. To highlight the concerns identified, it will consider recent engagement by various courts with GenAI use. In the absence of any (current) decision directly involving a valuer, this consideration will engage with decisions involving self-represented litigants (those without legal representation) and lawyers. It concludes by providing thoughts for the future.

Issues for valuers

As was anticipated, many 'human tasks' are now automated or are being replaced by the use of machines (PWC, 2018). In valuation practice, this has already led to the automation of some valuation processes, for example, the mass appraisals seen for statutory land valuation purposes, i.e., rating and land taxation (Hoang et al., 2024). Despite the potential uses of GenAI identified by other authors (Cheung, 2024a), due to the legal and other risks associated, it is submitted that not all uses, and in any event no unthinking uses, are appropriate for valuation professionals.

AI use can lead to greater accuracy in valuation reports (Abidoye, 2017), for example, by using approved templates, in the ease of comparing and analysing data, and for checking for spelling and grammar accuracy. Nevertheless, there remains the need to ensure that professional standards are adhered to in the preparation and presentation of a valuation report. In many countries, professional valuers (howsoever called) are required to comply with the *International Valuation Standards* ('IVSC'), as well as local professional Codes of Practice. Relevant IVSC obligations include, but are not limited to, the requirements to apply the methodology appropriate to the asset or interest being valued (IVS 103) and, where the interest being valued is, or relates to, real property (i.e., land), to inspect the property and engage with all searches and other relevant information '*to ensure that the* valuation *is properly supported*' (IVS 400).

Tasks such as the need for an actual inspection of the property, the critical engagement with search and other data, and simply bringing years of experience to the valuation process remain very much human-centred (Abidoye, 2017). Therefore, there remains a need for active engagement by the individual valuer whose name and signature will appear on the report. This need is reinforced by the information required to be contained in a valuation report (IVS 106). It is also important to remember that it is the named valuer who will be held responsible for the contents of the report (Cradduck, 2015a; Wilson, 2007). One concern about the use of ChatGPT and other similar services is that the outputs produced can include fabricated data and sources, and can misattribute content to the wrong real source. Absent skills in AI prompt generation (Cheung, 2024a), a valuer is at risk of their report being inaccurate (Cheung, 2024b). If there are inaccuracies in either the analysis or in the conclusions reached, even if in some part engaging with accurate data, these will taint the whole report, making it unreliable (Emsley, 2023).

Clients' understandings and engagements with the valuation process, and with the valuation report, will vary depending upon what 'type' of client they are and their level of sophistication (Levy and Schuck, 1999). However, in one respect, all clients are the same – they will seek to rely upon the valuation provided, and if it is inaccurate and they suffer a loss as a result (and sometimes even when any loss is not yet a possibility – see Cradduck, 2015b), they will seek to hold the named valuer liable (Wilson, 2007). Unfortunately, due to the common use of disclaimers in most software licences and agreements, if a valuer is found to be negligent due to the incorrect analysis of data produced by a GenAI or other type of software tool, most likely they will not be able to shift their liability to the software's creator.

Relevantly for this discussion, and separate from concerns about the accuracy of any output, any use of any software tool for the preparation of the valuation report also leads to considerations of, and the requirement to comply with, obligations relevant to both privacy and data security.[1] The use of any proprietary (pay-to-use) software tool is regulated by the terms of the user's agreement with the software provider. Such agreements include provisions relating to the ownership, use, and management of the software and related IP (Marotta-Wurgler, 2007; Todewale, 2022). Examples include the number of licences granted; the length of the agreement; ownership of the software IP rights (usually the provider); ownership of the IP rights in the data entered (usually the user, or their client); ownership of the IP rights in any output (it depends); and what use/s may be made of the data (if any) by the software provider. These also commonly provide for the management of liability issues and establish mechanisms to address privacy and confidentiality concerns and for managing other disputes (Marotta-Wurgler, 2007; Todewale, 2022). Conversely, while open-access software tools have the benefit of being (often considerably) less expensive, these are unlikely to have the same protections of data as proprietary software (Zhao et al., 2021). Therefore, even when the content of the report raises no concerns, issues may arise from the need to input client-provided, or client-related, information to generate the analysis and or create the ultimate report.

Therefore, while GenAI tools *may* assist in the preparation of reports, these tools should never be used unthinkingly. Concerns exist regarding both the accuracy of any analysis and the conclusion of the outputs created, <u>and</u> the security of any data entered in any GenAI system. Any inaccuracies or incompleteness, or other failings in the resulting report, will also lead to the courts not accepting either the written or oral evidence being presented. In turn, an unthinking use also may lead to disciplinary actions as well as liability concerns both for individual valuers and valuation firms.

Courts' considerations of GenAI use

In the absence of decisions involving valuers, examples of the conduct of self-represented individuals and lawyers will be considered to illustrate the courts' concerns about the accuracy of GenAI-produced data. This concern, regrettably, is seen most obviously in recent matters when ChatGPT has been used as a 'shortcut' to proper legal research.

In 2023, in a UK matter involving an appeal against a liability arising due to an unreported capital gains tax, a self-represented defendant, Mrs Harber, used ChatGPT to assist with undertaking 'legal research'. This 'legal research' resulted in her providing the details of nine created cases to the Tribunal in support of her application. In considering her appeal against a penalty of £3,265.11, the Tribunal opined that '*citing invented judgements is* [not] *harmless. It causes the Tribunal and HMRC to waste time and public money, and this reduces the resources available to progress the cases of other court users who are waiting for their appeals to be determined*'. (*Harber*, 2023 [24]).[2]

More recently, in 2024, in another UK decision involving a self-represented litigant, Mr Oakley, who was seeking *freedom of information* access to documents, the Tribunal was required to consider the reliability (or not) of GenAI-created outputs to determine the weight that should be given to an analysis undertaken purely using ChatGPT. The result was that the Tribunal gave little weight to such an output because there was

> *no evidence … as to the sources the AI tool considers when finalising its response nor is the methodology used by the AI tool explained. <u>If comparisons are drawn to expert evidence, an expert would be required to explain their expertise, the sources that they rely upon and the methodology that they applied before weight was given to such expert evidence</u>.*
>
> [emphasis added] (Oakley, 2024, [27])[3]

In the context of conduct that may lead to disciplinary action, the consideration of US District Judge Hon. P. Kevin Castel of June 2023 provides an internationally recognised example of the consequences of inappropriate engagements by a legal professional with GenAI content, the consequences for which were compounded by the lawyers not being honest about their conduct at the earliest opportunity. District Judge Castel observed that:

> In researching and drafting court submissions, good lawyers appropriately obtain assistance from junior lawyers, law students, contract lawyers, legal encyclopedias and databases such as Westlaw and LexisNexis. Technological advances are commonplace and there is nothing inherently improper about using a reliable artificial intelligence tool for assistance. But existing rules impose a gatekeeping role on attorneys to ensure the accuracy of their filings.
>
> [emphasis added] (Mata, 2023)

Mata highlights the responsibility placed upon the person signing any document. In this context, 'Mr. LoDuca signed the Affirmation in Opposition … [but] was not its author' and 'simply relied on a belief that work produced by Mr. Schwartz, a colleague of more than twenty-five years, would be reliable' (Mata, 2023, [6]). Mr LoDuca was wrong. As Judge Castel also noted in his preliminary comments, instead of using reliable legal databases, the lawyers had 'abandoned their responsibilities when they submitted non-existent judicial opinions with fake quotes and citations created by the artificial intelligence tool ChatGPT, then continued to stand by the fake opinions after judicial orders called their existence into question'. (Mata, 2023) Unfortunately, apparently Mr Schwartz had not appreciated that ChatGPT was not a reliable database nor that it could fabricate cases (Mata, 2023, [11], [37]). The lawyers each had a 'bad faith' finding made against them (Mata, 2023 [23], [24];[4] their law firm was held responsible for their conduct (Mata, 2023 [25])); they had to write to the plaintiff, and write to each Judge cited as the author of a fake case; and the lawyers and firm were jointly fined $5,000 (Mata, 2023).

Mata (2023) was considered in 2024 by Humphreys J of the Victorian Supreme Court. An Australian legal practitioner had provided a summary of authorities to the Court generated by AI and not checked by the legal practitioner (*Dayal*, 2024). The legal practitioner *'did not fully understand how the* [AI] *research tool worked…* [but] *acknowledged the need to verify AI assisted research, or indeed any source of legal research relied upon, for accuracy and integrity'* (*Dayal*, 2024, [8]). Humphreys J observed *Mata* (2023) had drawn *'worldwide attention to the risk of relying on generative AI for research purposes in litigation without independent verification'* (*Dayal*, 2024, [11]). In referring the practitioner to the Victorian Legal Services Board (*Dayal*, 2024, [22]), Her Honour opined, '*it is in the public interest for the* [VLSB] *and Commissioner to be aware of the professional conduct issues arising in this matter, given the increasing use of AI tools by legal practitioners in litigation more generally'*. (*Dayal*, 2024, [21]).

In December 2024, Judge Skaros of the Federal Circuit and Family Court of Australia had to engage with the written submissions of a senior legal practitioner (*Valu*, 2025, [26]), which purported to rely on 17 decisions, none of which existed, and portions of the Administrative Appeals Tribunal's prior decision in the matter, which also did not exist (*Valu*, 2025, [1], [4]). The legal practitioner had '*decided to use AI, which had been promoted by reputable legal services such as LEAP and Lexis Nexis as being of assistance in legal practices*'; however, he had not used those providers' services but instead used ChatGPT (*Valu*, 2025 [22]).

In making his ruling, Skaros J opined that:

> ... filing an application and submissions which contained citations to ... cases which do not exist and alleged quotes from the Tribunal's decision which do not exist, falls short of the standard of competence and diligence that the applicant in the substantive proceedings was entitled to expect from his legal representative. The conduct also falls short of a legal practitioner's duty to the Court, including the duty to ensure that the Court is not deceived or mislead, even if unintentionally...
>
> (*Valu*, 2025, [18])

Accordingly, Skaros J also held it was 'in the public interest' to refer the legal representative to the NSW Legal Services Commissioner for consideration of his conduct (*Valu*, 2025 [38]).[5]

Discussion

Courts often rely on valuation reports and valuers as expert witnesses to determine the value of an asset or an interest. As Crosby et al. (2003) articulated, '*the expert witness's primary duty lies to the court and not to the client. The witness is expected to provide honest, impartial and unbiased evidence...*' (Crosby et al., 2003, p.282). It is crucial, therefore, that the Court or Tribunal can have confidence in the expertise of such experts, in the written reports provided by those experts, and in the oral evidence they give based on their written reports.

As the cases discussed above highlight, using GenAI will often result in outputs that are fabricated and misleading. In response, courts are beginning to proactively manage for the use of AI, for example, by drawing to practitioners' attention the need to ensure that those who use it are skilled in its use and warning that those tools are not engaging in legal research.[6]

However, in other instances, courts are prohibiting the filing of documents created using GenAI. This is seen, for example, and as noted in *Valu* (2025, [23]), in the *Practice Note* issued by Chief Justice Bell of the Supreme Court of New South Wales, which applies to all litigation matters (*PN SC GenAI*, 2025).[7] Relevantly for valuers, this provides:

> 19. Expert reports are required to state the opinion or opinions of the expert, and his or her reasoning process.
>
> 20. ... Gen AI must not be used to draft or prepare the content of an expert report (or any part of an expert report) <u>without prior leave of the Court</u>.
>
> (emphasis added)

The relevance of the identified issues and concerns for the valuation profession can be summarised as the requirement to be conscious of four needs:

1 the need to ensure that data engaged with is from a reputable source;
2 the need to ensure all employees are trained in AI use – both as to how AI can support their practice and what are, and what are not, reputable systems;
3 the need to be aware of any prohibition on the use of GenAI; and
4 the need to ensure valuers remember the valuer/s whose name/s and signature/s appear on the valuation report will be held responsible for the contents of that report.

The Practice Note's prohibitions and the observations of Humphreys J[8] and Skaros J,[9] are reminders of the risks involved in any GenAI use. For employers, these also highlight the need for vigilance in the training and monitoring of employees, no matter how experienced.

Conclusion and thoughts for the future

To identify that a matter is an issue or a concern requiring further consideration or investigation, it is necessary to be able to identify when something does not look 'right'. That is, something is odd, or uncommon or not expected. The more AI is used, the more common it becomes, and the less obvious (perhaps) the errors in outputs. This may be appropriate when simple spelling needs to be corrected, but when changes are made without the human specifically considering and adopting each change, then the reliability of the information is questionable.

This chapter has highlighted but a few of the current and potential legal risks specifically facing the valuation profession in its use of GenAI, with a focus on the reports produced and the role of an expert witness. Relying on any material not actually created by the user is always fraught with risk. It is now also important to remember that whenever engaging with documents and materials not created directly and personally by *you*, identifying what is the product of a machine and what is the product of an individual human will become even harder. When a document is produced by a human using GenAI, then there is the added risk that some (if not all) sources and content will be fabricated and/or that real data will not be correctly analysed.

When a valuation report is finalised, whatever its future use may be, it is required to be signed[10] by the valuer. As with any valuation report, the valuer is responsible for what the report contains. Where the report is produced in part by GenAI, the overarching consideration for the valuer who signs it is that, if the AI is wrong, it is not the software development company that will be sued and disciplined; it will be that valuer. The bigger, the better; the more ubiquitous AI use becomes, then (perhaps) the better the world will be for all. The more experienced an individual is at using GenAI then, and unlike the

AI-inexperienced lawyers in the cases discussed, the less likely that valuer is to take any document at face value.

The ultimate lesson for any engagement with GenAI is quite simple. *Just because you can, does not mean you (unthinkingly) should.*

Questions

1 How do the courts view the use of generative AI tools?
2 How reliable are generative AI tools for legal research?
3 Are generative AI tools a substitute for legal (or other) research?

Notes

1 For example, see the obligations in the European Union's *GDPR*, which has internal and external application. (Cradduck et al., 2021). This applies in addition to specific regulations under the European Union's *AI Act*.
2 Noting that Mrs Haber's appeal failed (*Harber*, 2023, [62]).
3 Noting that Mr Oakley's appeal was dismissed (*Oakley*, 2024, [40]).
4 Other breaches related to the lawyers' engagements with the Court, Affidavits made, and evidence given during the sanctions hearing. These are beyond the scope of this Chapter to consider.
5 Noting also that there were other concerns relating to the legal practitioner's engagements with the Court. As of the date of finalisation of this Chapter, I have not been able to locate any disciplinary action by the LSC.
6 See discussion in *Dayal* (2024) at [14]–[15].
7 PN SC GenAI (2025) applies to all court proceedings from 3 February 2025.
8 *Dayal*, 2024, [10]: '…the use of AI tools offer opportunities [but] also comes with significant risks'.
9 *Valu*, 2025, [37]: '… the misuse of generative AI is likely to be of increasing concern…'.
10 Either by doing so in hardcopy, by pasting their signature, or by affixing a digital signature.

References

Abidoye, Rotimi, and Chan, Albert (2017). Valuers' receptiveness to the application of artificial intelligence in property valuation. *Pacific Rim Property Research Journal, 23*(2), 175–193.

Baecker, Ronald (Ed.). (2014). *Readings in Human-Computer Interaction: Toward the year 2000*. Elsevier.

Baron, Rachel (2024). AI editing: Are we there yet? *Science Editor*, 47(3), 78–82.

Cheung, Ka (2024a). Real estate insights unleashing the potential of ChatGPT in property valuation reports: The 'Red Book' compliance Chain-of-thought (CoT) prompt engineering. *Journal of Property Investment & Finance*, *42*(2), 200–206.

Cheung, Ka (2024b). Real estate insights: Establishing transparency–setting AI standards in property valuation. *Journal of Property Investment & Finance*, *42*(4), 406–408.

Cradduck, Lucy (2015a). Professional indemnity insurance, performance and pre-emptive negligence: The impact of 'non-claims'. In Bond, S (Ed.) *Proceedings of the Pacific Rim Real Estate Society 21st Annual Conference*, 1–6.

Cradduck, Lucy (2015b). The statutory defence(s) of 'peer professional opinion'. *Australia and New Zealand Property Journal. 5*(3), 224–228.

Cradduck, Lucy, Stevens, Scarlett, and Cowan, Matthew (2021). Data sharing, international property practices and the GDPR: Communicating with your consumers. *Property Management*, *39*(1), 22–33.

Crosby, Neil, Kincaid, Diane, Murdoch, John, and Lavers, Anthony (2003). Expert valuation witnesses in Australia and the UK. *Journal of Property Research*, *20*(3), 281–304.

Dale, Robert, and Viethen, Jette (2021). The automated writing assistance landscape in 2021. *Natural Language Engineering*, *27*(4), 511–518.

Dayal (2024) FedCFamC2F 1166.

Emsley, Robin. (2023). ChatGPT: these are not hallucinations – they're fabrications and falsifications. *Schizophr 9*, 52. https://doi.org/10.1038/s41537-023-00379-4

Harber v Commissioners for His Majesty's Revenue and Customs (2023) UKFTT 1007 (TC) ('Harber')

Hoang, Viet Ngu, Nguyen, Kien, Nguyen, Manh Thang, and Blake, Andrea (2024). Unlocking visual data to enhance the accuracy of AI-enabled mass valuation of urban houses: An Australian city case study. *Expert Systems with Applications*, *255*(Part D), Article number: 124784. https://doi.org/10.1016/j.eswa.2024.124784

Kyosev, Yordan, and Boussu, Francois (Eds.) (2022). *Advanced Weaving Technology*. Springer Nature, Switzerland AG.

Levy, Deborah, and Schuck, Edward. (1999). The influence of clients on valuations. *Journal of Property Investment & Finance*, *17*(4), 380–400.

Marotta-Wurgler, Florencia (2007). What's in a standard form contract? An empirical analysis of software licence agreements. *Journal of Empirical Legal Studies*, *4*(4), 677–713.

Mata v. Avianca, Inc., 678 F. Supp. 3d 443 (S.D.N.Y. 2023). Opinion and order on sanctions. ('Mata').

Murugesan, Sam (2025). The Turing test at 75: Its legacy and future prospects. *IEEE Intelligent Systems*, *40*(1), 20–24.

Nardinelli, Clark (1986). Technology and unemployment: the case of the handloom weavers. *Southern Economic Journal*, *53*(1), 87–94.

Oakley v Information Commissioner [2024] UKFTT 315 (GRC) (18 April 2024) ('*Oakley*').

Pavlić, Ana (2024). Endangered gender: Her between learning and teaching. In Silvana Carotenuto, Maša Huzjak, Renata Jambrešić Kirin, and Biljana Kašić (Eds), *Feminist Trans/Formations: Media, Art, Literature*, (UniorPress, Napoli), 5–49

Pigozzi, Federico (2023). 'Of Typewriters and PCs.' *Proceedings of the ALIFE 2023: Ghost in the Machine: Proceedings of the 2023 Artificial Life Conference. ALIFE 2023: Ghost in the Machine: Proceedings of the 2023 Artificial Life Conference*. July 24–28, 2023. Online. (pp. 141). ASME. https://doi.org/10.1162/isal_a_00693

PWC. (2018). Workforce of the future: The competing forces shaping 2030. https://www.pwc.com/gx/en/services/people-organisation/workforce-of-the-future/workforce-of-the-future-the-competing-forces-shaping-2030-pwc.pdf (accessed on 18/02/2025).

Regulation (EU) 2016/679 of The European Parliament and of The Council of 27 April 2016 on the protection of natural persons with regard to the processing of personal data and on the free movement of such data, and repealing Directive 95/46/EC (General Data Protection Regulation) ('*GDPR*')

Regulation (EU) 2024/1689 of The European Parliament and of The Council of 13 June 2024 laying down harmonised rules on artificial intelligence and amending Regulations (EC) No 300/2008, (EU) No 167/2013, (EU) No 168/2013, (EU) 2018/858, (EU) 2018/1139 and (EU) 2019/2144 and Directives 2014/90/EU, (EU) 2016/797 and (EU) 2020/1828 (Artificial Intelligence Act) ('*AI Act*')

Supreme Court Practice Note SC Gen 23 – use of Generative Artificial Intelligence (Gen AI), Version 28 January 2025, https://supremecourt.nsw.gov.au/practice-procedure/generative-artificial-intelligence.html ('*PN SC GenAI*')

Todewale, Priyanka. (2022). Key contractual provisions of an end user license agreement. *Indian Journal of Law and Legal Research*, *4*, 1–7.

Turing, Alan (1950). Computing machinery and intelligence. *MIND, A Quarterly Review of Psychology and Philosophy*, *59*(236), 433–460. https://phil415.pbworks.com/f/TuringComputing.pdf

Valu v Minister for Immigration and Multicultural Affairs (No 2) [2025] FedCFamC2G 95 ('Valu')

Wilson, T (2007). Valuers' and agents' liability issues. *Australian Property Law Bulletin*, *21*(8), 81–86.

Zhao, Yuhand, Liang, Ruigang, Chen, Xiang, and Zou, Jing (2021). Evaluation indicators for open-source software: A review. *Cybersecurity*, *4*, 20. https://doi.org/10.1186/s42400-021-00084-8

Chapter 12

Reflection on AI, big data and analytics in real estate valuation

Simon Benjamin Fonteyn

Introduction

The way we value real estate is changing – fast. Gone are the days when I started out in the property valuation industry, where we relied on a mix of gut feelings, experience, formal and informal property networks and historical and current sales and leasing data.

Today, with the rise of AI, which refers to computational systems designed to perform tasks that simulate human cognitive functions, including learning, reasoning, pattern recognition and decision-making, using algorithms, statistical models and large-scale data processing. We now have access to real-time analytics, predictive modelling and automated insights that are reshaping the future of the industry. While these shifts bring incredible opportunities, they also raise questions: What role will valuation professionals play in a world increasingly dominated by algorithms? Can AI really value heterogeneous and complex properties more accurately than valuation professionals? What skills will the valuer of the future need, and what might future valuation reports incorporate? How is the property profession responding to the adoption of AI and Big Data?

This reflection piece explores how AI and Big Data, which are defined as extremely large, complex, and high-velocity datasets that exceed the processing capacity of traditional data management tools, are transforming valuation practices. The advantages and limitations of AI-driven models are also explored. I have examined Australian case studies of the adoption of AI and Big Data and interviewed two leading researchers in AI and Property Technology ('Proptech') to gauge their insights.

Moving beyond traditional valuation methods

Traditionally, valuation professionals like me would visit a site, assess its physical, locational and financial characteristics, compare it to recent sales and lettings, and apply professional judgement to arrive at a value. This process was reliable but also inherently backwards-looking. It was also extremely

DOI: 10.1201/9781003539254-16

time-consuming and often involved manual incorporation of multiple data points, including sales, lettings, planning controls, geospatial data, environmental factors, population growth and economic indicators, delivered via a static PDF report. Often, the required information would be outdated or not publicly available, and we would need to rely on our professional networks to bridge those data gaps.

AI and Big Data have started to change this by incorporating a broader range of qualitative and quantitative factors, including economic indicators and sales data, and letting data, both historical and on the market, to facilitate more informed, real-time assessments. However, the reliance on real-time data, such as 'on the market' listings, must be contextualised, as these often reflect aspirational pricing rather than market value.

One of the biggest advantages AI offers is in pattern recognition (though not always conclusive). Machine learning algorithms process massive datasets through the big data process, identifying correlations that might not be obvious to a human valuer. For instance, AI can examine the correlation between changes in truck volumes along arterial routes in and out of industrial land corridors as a forward predictor of industrial rents and capital values.

Geospatial analytics and computer vision

Location has always been a critical element of property valuation; however, AI and geospatial analytics have taken this to a whole new level. AI can analyse mass mobility data, computer-generated traffic flows, and even foot traffic data from cameras to understand how the dynamics of a location function in real life.

For example, data from mobile devices can reveal how many people pass by a retail property at different times of the day and night, assess their dwell time and analyse cross-visitations, giving a much clearer picture of a site's value to a retail business. AI can also analyse satellite imagery to detect urban development trends, environmental risks, car park utilisation and neighbourhood gentrification patterns. This ability to incorporate complex, multi-layered datasets can produce more dynamic and accurate property assessments.

Another emerging and particularly powerful application of AI in valuation is computer vision. For example, consider the Australian PropTech company, Archistar, that developed a visualisation platform integrating AI and computer vision to enhance the analysis of urban development potential. By processing zoning regulations, aerial imagery and architectural parameters, this platform enables the rapid generation of compliant 3D building models. AI algorithms evaluate site constraints and development controls, while computer vision facilitates the transformation of geospatial and design data into visual simulations. This technology supports the modelling of development feasibility scenarios by allowing users to test various built form outcomes and optimise land use. Additionally, it contributes to preliminary development valuations and feasibility studies by estimating potential yield and density, thereby informing strategic planning and investment decisions.

AI-powered image and video analysis can assess property conditions by processing thousands of photos and videos, both current and historical. These models can assist in identifying a range of issues for valuation professionals, including building measurement reconciliations, identifying dates of construction and architectural heritage, detecting structural weaknesses, identifying defects and even pinpointing environmental hazards such as asbestos materials, lead-based paint, pest infestations or water damage.

However, Dr Biyanka Ekinayake of the University of Technology in Sydney cautions against overreliance on this technology.

Determining the underlying cause of a defect often requires more than visual/textual input, such as structural history, material properties or environmental conditions, which are not always captured in the input data. Specifically, AI-driven vision-language models often function as 'black boxes', providing limited transparency in how conclusions are reached. Vision-language models can misinterpret contextual cues or make incorrect assumptions if the textual input is vague or ambiguous or leading. A well-annotated, diverse training dataset with clear, high-resolution images and text information is essential to train the AI model to accurately detect defect types, interpret causes and provide meaningful recommendations. Therefore, the accuracy and reliability of AI results are highly dependent on the quality of the input images and the robustness of the training process. Overreliance on AI predictions without human oversight could lead to incorrect diagnoses or unsuitable remediation strategies'. Dr Ekinayake sends a prescient message that such analyses should not replace expert engineering assessments.

Risk assessment is another area where AI is being utilised in real estate. Traditionally, landlords, investors and lenders relied heavily on historical performance and market sentiment to assess risk. Today, AI can analyse a broader range of forward-looking indicators – rental arrears, tenant payment patterns, digital twin data and economic trends – to forecast potential tenant defaults, asset lifecycle triggers or even broader macroeconomic downturns. Imagine an AI system that alerts a shopping centre owner when a tenant's sales are in decline, indicating a likely lease termination months in advance. An example of such technology is Credia AI by Re-Leased, which integrates predictive analytics with real-time property and financial data to flag risk exposures and automate decision-making. Credia AI accelerates routine administrative tasks, thereby reducing processing time. It also provides property professionals with strategic foresight by analysing tenant behaviour, market fluctuations and portfolio health. This allows for a more proactive approach to asset and risk management, helping reduce exposure to financial uncertainty and improving investment outcomes and tenancy mix planning.

As platforms like Credia AI continue to generate increasingly granular and predictive data on tenant performance, leasing risk and asset condition, valuation professionals may be expected to incorporate such forward-looking intelligence into their valuation models, especially for complex income-producing assets. This will shift away from static, historical valuation inputs towards more dynamic, risk-adjusted assessment methodologies.

AI is also highly valuable in complex property valuations that involve forward-looking estimates, such as Discounted Cash Flow (DCF) modelling. Traditionally, DCF calculations required manual input of assumptions regarding rental growth, capitalisation rates, inflation and other risk factors to predict terminal or ending values for these models, which can represent a considerable proportion of the total value of the asset.

One of the key benefits of AI in DCF modelling is its ability to run multiple predictive scenarios based on different economic conditions, tenant behaviours, asset life cycles and market trends. AI can assess how changes in interest rates, supply–demand imbalances or regulatory shifts impact key financial metrics such as rents, capitalisation rates, Internal Rates of Return (IRR), and capital values into the future. By leveraging machine learning, AI can identify patterns and correlations that traditional valuation models might overlook, resulting in a more informed and proactive valuation approach that caters for multiple scenarios. However, given that terminal values can represent a large portion of overall values, human oversight is still critical. The professional standard bodies of valuation and the professional insurance market will also need to review and ultimately regulate the use of AI in DCF modelling.

AI has a growing role in PropTech and property valuation

For this reflection chapter, I interviewed Dr Stephen Friel, Managing Director of Red Beard Consulting, who advises high-growth PropTech companies, deploys large-scale data infrastructure and develops predictive AI models that extract alpha in asset pricing – whether in equities, infrastructure or increasingly, real estate. Dr Friel contributed to one of the first forward predictors of residential real estate markets in Australia.

The Australian property valuation industry presents a compelling case study on how AI and data-driven methods are redefining a traditionally analogue market. Total Addressable Market (TAM) analysis shows that beyond core residential mortgage valuations – which dominate current AVM use – Australia's fragmented asset base includes aged care facilities, strata-titled buildings, data centres, medical precincts, retail and commercial properties and new developments. Each of these presents a unique market segment with an estimated combined TAM exceeding $700 million annually in valuation and adjacent services. However, much of this remains unaddressed or under-serviced due to structural limitations in the traditional valuation delivery model: manual workflows, inconsistent data standards and low frequency of serviceable updates.

Technologically, this is now solvable. We have seen AI and machine learning models deliver outsized value in capital markets – from automated company valuation frameworks using earnings and alternative data to natural language processing (NLP) engines mining public filings for sentiment signals. The same principles – data fusion, predictive analytics and risk-weighted model outputs – are being applied to physical assets. In the valuation of real property, these tools are increasingly becoming available. AVMs are already widely used for residential valuations. PropTrack, CoreLogic, PEXA and others are leveraging machine vision, dynamic regression and even buyer sentiment tracking to refine model confidence and extend coverage. But it is the emerging and non-residential asset classes where the greatest disruption and opportunity now lies'.

Case study – Value Australia Project

One significant case study illustrating the application of AI in the Australian valuation market is the Value Australia Project, now a PEXA Group company.

Value Australia is a pioneering project that combines AI, machine learning and big data to develop a national platform for automated and scalable property valuations. It is led by FrontierSI in collaboration with government, academia, and industry partners, including the NSW Valuer General, the Commonwealth Bank and multiple state governments. The objectives are:

- To automate land and property valuation across Australia.
- To support government decision-making in taxation, infrastructure and planning.
- To enable fairer, faster and more accurate valuations.
- To create spatial digital twins and simulate urban development scenarios.

Technology and methodology

The technology and methodology utilised are as follows:

- AI and Machine Learning
    - Trained on large volumes of property transaction data, property attributes, environmental data and spatial analytics.
    - Used to predict values at both individual and area levels with explainable models.

- Spatial Digital Twins
    - Integration of property valuation models with 3D city models and GIS systems.
    - Enables users to visualise the impact of planning or zoning changes on property values.

- Explainable AI (XAI)
  - Ensures transparency in how valuation decisions are made, crucial for government use and regulatory compliance.

Key outcomes and recognitions

- Awards for Innovation: Value Australia received the 2023 Cooperative Research Australia Award for Research Commercialisation and the Asia-Pacific Spatial Excellence Award for Innovation, recognising its successful transition from research to a commercial product.
- Pilot Programmes and Accuracy: The platform was piloted in metropolitan Sydney and Brisbane as part of the RAISE project, achieving alignment with market sales data to 94% accuracy.
- Commercial Adoption: Value Australia has secured a multi-year deal with a Tier 1 bank and initiated a pilot programme with the NSW State Government, indicating its practical application and industry acceptance.

In 2024, Value Australia signed a Memorandum of Understanding (MoU) with the Australian Property Institute (API) to promote the ethical use of AI in property valuation and support continued innovation in digital valuation standards. This partnership reflects growing institutional support for AI-enabled valuation models that maintain transparency and professional integrity. The MoU with the API emphasises collaboration on research, skills development and policy to ensure valuation professionals remain central to AI adoption and that technology augments, rather than replaces, their role in the profession. The Value Australia Project represents a landmark example of how AI and spatial data science can modernise property valuations at scale. It highlights the potential for intelligent, real-time valuation systems to improve the accuracy with which various levels of government assess property-based taxation revenue, while maintaining market transparency. It sets a precedent for AI adoption in public-sector property and land assessment and serves as a model for global replication.

The future of property valuations: Humans and AI working together

As AI continues to evolve, many in the industry wonder whether valuation professionals will be replaced by AI. The short answer? Not anytime soon. While AI excels at data processing and pattern recognition (though not always conclusive), human expertise remains crucial for interpretation, judgement, bridging market data gaps and ethical oversight.

AI-generated reports can provide deep insights, but they still need a professional's eye to validate results, account for qualitative factors and ensure compliance with industry standards. Moreover, valuation is not just about crunching numbers – it is about understanding market sentiment, understanding complex deals and providing strategic advice to clients. These are areas where property professionals will be needed. Lastly, it remains to be seen whether AI could ever be considered an expert and therefore could be used in court cases, as AI is not currently admissible as expert evidence, as it cannot be cross-examined under legal standards of expert evidence. Recently, the NSW Supreme Court has banned all AI-generated reports for compiling expert evidence, citing transparency and reliability concerns.

Another challenge AI faces is the quality and transparency of data. AI models depend on large datasets, which are continually updated; without them, biases can lead to inaccurate valuations. For instance, if certain types of properties or locations are under-represented in the training data, AI models struggle to provide fair and balanced assessments. Many key data inputs, such as rental data and incentives, are subject to confidentiality agreements or the Privacy Act and, therefore, not discoverable by AI. This limitation means that AI models may work with incomplete datasets, reducing their accuracy and reliability of automated valuations.

There is also growing concern that some early-career valuation professionals may over-rely on AI-generated outputs without possessing the requisite expertise to critically assess their validity. To safeguard the integrity of the profession, it is essential that future valuers are comprehensively trained in both traditional valuation methodologies and emerging digital technologies.

Increasing adoption of automated valuation models

The adoption of Automated Valuation Models (AVMs) by major Australian banks has accelerated over the past decade, driven by advances in data analytics and the need for faster, cost-effective lending decisions. Institutions such as Commonwealth Bank, Westpac and ANZ now routinely use AVMs for low-risk residential mortgage lending, particularly for loans with low loan-to-value ratios (CoreLogic, 2022a). These models utilise large datasets, including recent sales and property characteristics, to produce rapid valuations that align with internal credit risk thresholds. Regulatory bodies such as the Australian Prudential Regulation Authority (APRA) have acknowledged AVMs as part of the banks' broader valuation frameworks, provided appropriate risk controls are in place (APRA, 2018a). However, concerns remain regarding their accuracy in heterogeneous or thinly traded markets, prompting continued reliance on human valuation for complex or high-risk assets (RICS, 2021).

The skill set of the future valuer and future valuation reports

The skill sets required of tomorrow's property valuer will look different from what the profession has traditionally demanded. Where valuation professionals once focused primarily on manual analysis, the future will call for a more interdisciplinary approach – one that blends real estate knowledge with technological fluency.

To stay relevant, valuation professionals will need to collaborate more closely with data analysts and researchers, providing strategic oversight for the models being built and the outputs they produce. They will also need a working knowledge of emerging technologies such as geospatial analytics, computer vision and the fundamentals of machine learning. This shift will require a rethinking of how we educate the next generation of valuation and property professionals.

Valuation reports themselves may also evolve, and we may expect traditional static PDF documents to be augmented by richer, more dynamic formats. Future reports may include:

- Video footage of the subject property and comparable sales
- AI-driven geospatial and mobility data as part of mainstream analysis
- An autonomous, automated model-generated value opinion

These reports may no longer even be called valuations. The insurance industry may also require different professional indemnity risk premiums for human-generated valuations vs AVMs. Lastly, the latter may not even be provided by traditional valuation firms. For example, big data companies such as PEXA and Core Logic now offer AVMs for residential markets to their clients. These reports may be delivered via chatbot interfaces and online portals, tailored to specific end users, such as board-ready financial reporting formats.

Responsible use of AI in real estate valuation: Reflections on the new RICS professional standard

In 2024–2025, RICS released its first professional standard on the responsible use of AI in valuation, and it could not have come at a more relevant time. As someone working in real estate valuation and education, I have seen firsthand how AI tools – particularly those for modelling and analytics – are becoming more embedded in our day-to-day work. This standard really zeroes in on the risks we all need to be mindful of, like bias in outputs, unreliable data, and the 'black box' nature of many systems. What I appreciate is that it does not just tell us to be cautious; it lays out a practical framework. We are expected to have a working knowledge of how AI functions, maintain a risk register, evaluate whether AI is genuinely the right tool for the job, and keep clients informed whenever AI has been used in our work.

From its effective date in 2025, the standard sets clear expectations: we need to document our decisions, assess the sustainability of the tools we are using, and apply professional judgement rather than relying blindly on outputs.

What struck me most is the emphasis on transparency and explainability – clients have a right to know not just that AI was used, but how and why. And for those developing AI tools in-house, the responsibilities are even greater, covering everything from data ethics to stakeholder diversity. Overall, I see this standard as a timely and practical guide that helps bridge the gap between the potential of AI and the trust that underpins our profession.

The successful integration of AI into property valuation practice requires more than just technological capability; it demands a considered approach to change management. This includes addressing cultural resistance within firms, adapting professional workflows and ensuring alignment with regulatory and ethical standards. Many valuation practitioners – particularly those accustomed to traditional methods – may be sceptical of AI tools due to concerns about transparency, professional liability, or the perceived erosion of expertise and, in particular, the soft skills required to handle clients, market professional services and deal with conflicts and disputes.

To facilitate adoption, leadership within firms must promote a mindset of continuous learning, provide structured training on AI applications, and foster collaboration between technical and valuation teams. Additionally, industry bodies and institutions must develop clear frameworks and guidelines to support the responsible implementation of AI, reinforcing confidence in its use while maintaining the core values of the profession.

Conclusion

Big Data and AI are fundamentally reshaping the real estate valuation landscape. The shift from historical analysis to predictive modelling enhances accuracy, efficiency and risk management. Geospatial analytics, machine learning and computer vision allow for deeper property assessments with increasing precision in residential homogeneous markets. However, caution is still required, as their overall reliability is questionable for heterogeneous properties, as well as complex assets and thinly traded markets and therefore requires human oversight.

Despite these advancements, human expertise remains irreplaceable. AI is an incredibly powerful tool, but its insights are most valuable when combined with professional judgement, experience and ethical considerations. Rather than replacing valuation professionals, AI enhances their capabilities, enabling us to work faster, more accurately and with greater strategic foresight.

Lastly, Dr Friel's insights highlight a pivotal moment for the property valuation industry. As AI and AVM technologies shift from theoretical potential to practical necessity, the profession, insurers and regulators must evolve – not to replace human expertise, but to amplify it. The future of valuation will belong to those who embrace data-driven tools while maintaining the rigour, judgement and governance standards that underpin trusted advice.

Discussion questions

1 How can professional valuers maintain ethical and professional standards in a valuation environment increasingly dominated by automated tools and predictive analytics?
   – Consider the role of transparency, human judgement and client communication in AI-integrated workflows.
2 What are the limitations of Automated Valuation Models (AVMs) in heterogeneous and complex property markets, and how might these limitations be addressed through regulatory or technological means?
   – Reflect on the quality of training data, model transparency and human oversight.
3 How might the valuation profession evolve to include interdisciplinary skills in data science and AI, and what changes would this require in education and professional accreditation?
   – Discuss potential curriculum updates, industry training, and continuing professional development (CPD).

Recommended further reading

CoreLogic (2022). Automated Valuation Models in Mortgage Lending: Methodologies, Challenges and Use Cases.

McKinsey Global Institute (2020). The Future of Work in the Age of AI.

Royal Institution of Chartered Surveyors (2025). *Responsible Use of Artificial Intelligence in Surveying Practice*. RICS.

A key international standard offering practical guidelines for integrating AI responsibly into valuation practices. https://www.rics.org/profession-standards/rics-standards-and-guidance/conduct-competence/rules-of-conduct

Offers insight into the practical application and risks of AVMs in the Australian residential lending sector. https://www.corelogic.com.au

Explores how AI and automation will transform job roles and skillsets across sectors, including property and finance. https://www.mckinsey.com/mgi

Bibliography

Archistar. (2024). Archistar's visualisation platform for urban development: AI and computer vision in planning. Retrieved from https://www.archistar.ai/

Australian Prudential Regulation Authority. (2018a). Information paper: Residential mortgage lending – Risk management expectations. https://www.apra.gov.au/sites/default/files/180504-Information-paper-residential-mortgage-lending-risk-management-expectations.pdf

Australian Prudential Regulation Authority (2018b). Referenced regarding regulatory acceptance of AVMs within the banking sector. Used in the 'Increasing Adoption on Automated Valuation Models' section. In-text citation: (Australian Prudential Regulation Authority, 2018).

CoreLogic. (2022a). Automated valuation models in mortgage lending: Methodologies, challenges and use cases. https://www.corelogic.com.au/

CoreLogic (2022b). Used to support trends in the adoption of automated valuation models (AVMs) by major banks. Mentioned in the 'Increasing Adoption on Automated Valuation Models' section. In-text citation: (CoreLogic, 2022).

Credia AI. (2024a). Credia AI by re-leased: Real-time analytics for commercial property intelligence. Retrieved from https://www.re-leased.com/credia-ai

Credia AI (2024b). Discussed in the 'Geospatial Analytics and Computer Vision' and Risk Assessment sections as a tool used in commercial property management. Provides predictive analytics for lease risk, tenant performance, and strategic portfolio insights. Cited as an example of how AI is being used for dynamic forecasting and automation in real estate. In-text citation: (Credia AI, 2024).

Dr Ekinayake, B. (2025a). *Personal interview conducted by the author*. University of Technology Sydney.

Dr Ekinayake, B. (2025b). Interview provided expert commentary on the limits of vision-language models in defect recognition. Mentioned in the 'Geospatial Analytics and Computer Vision' section. In-text citation: (Ekinayake, 2025).

Dr Friel, S. (2025a). Personal Interview Conducted by the Author, Who is the Director of Red Beard Consulting.

Dr Friel, S. (2025b). Interview elaborates on the disruption and opportunities AI brings to PropTech and asset pricing. Mentioned in the 'AI has a Growing Role in PropTech and Property Valuation' and Conclusion sections. In-text citation: (Friel, 2025).

OpenAI. (2025). ChatGPT (May 17 version) [Large language model]. https://chat.openai.com/. In-text citation: (OpenAI, 2025).

RICS (2021). *Automated valuation models: Roadmap for RICS members and stakeholders*. RICS. https://www.rics.org/content/dam/ricsglobal/documents/standards/rics_avm_roadmap.pdf

Royal Institution of Chartered Surveyors. (2025a). *Responsible use of artificial intelligence in surveying practice* (1st ed.). RICS. https://www.rics.org/profession-standards/rics-standards-and-guidance/conduct-competence/rules-of-conduct

Royal Institution of Chartered Surveyors (2025b). Cited in the discussion of responsible artificial intelligence use in valuation practice. Mentioned in the 'Responsible Use of AI in Real Estate Valuation' section. In-text citation: (Royal Institution of Chartered Surveyors, 2025).

Value Australia. (2024a). *Value Australia: AI, big data and digital twins for next-generation property valuation*. PEXA Group and FrontierSI. Retrieved from https://valueaustralia.com.au/

Value Australia (2024b). Detailed case study illustrating national AI-driven valuation implementation. Cited in the 'Value Australia Project' section for: – Explainable AI (XAI), 3D spatial digital twins, Pilot programs and commercial adoption, 2024 MoU with the API. In-text citation: (Value Australia, 2024).

Chapter 13

Quantum valuation

Navigating the disruption, seizing the opportunities in the new era

Greg Skulmoski, Alex Lang, Ashkan Memari and Cen Ying Lee

Why quantum matters in valuation

Quantum computing is poised to fundamentally transform property and asset valuation by enabling the rapid analysis of massive and complex datasets beyond the capabilities of classical computing. Unlike classical machines that process one calculation at a time, quantum computers use qubits, which allow for parallel computation at exponential speed. This capability opens new use cases in valuation, such as:

- High-frequency, real-time market forecasting.
- Valuing complex asset portfolios with interdependent variables.
- Risk optimisation for mergers, acquisitions, or development projects.
- Assessing asset sustainability using multi-source environmental data.

These use cases are made possible because quantum computers can solve problems that would take classical computers trillions of years—in a matter of hours. Thus, quantum valuation offers unprecedented gains in speed, accuracy, and analytical depth.

Property and asset valuers have long used technologies in their work processes; paper notebooks and ledgers gave way to early electronic spreadsheets like *Lotus 123* to complete complex calculations. We are guided by the Royal Institution of Chartered Surveyors (RICS) Red Book and its processes. Today, property and asset valuers use sophisticated software that may be in the cloud and buoyed with artificial intelligence to automate and optimise valuation processes like the income capitalisation approach. While the current promise of these technologies, like artificial intelligence, is alluring, think of AI (software) paired with quantum technologies (hardware) as truly transformational, with the biggest impact on business since the internet.

Expanding horizons: Additional applications in valuation

While we shared two valuation use cases based on quantum technologies (Figures 13.3 and 13.4), there are more. The AI use cases in this book can also

DOI: 10.1201/9781003539254-17

be tuned with quantum algorithms to run on quantum computers. The main advantage of quantum computing is that more big data can be accessed with quantum machine learning technologies, resulting in increased validity and reliability with reduced bias in property and asset valuations. Quantum computing and quantum algorithms can be applied to other property and asset valuations, including

1. Providing accurate market valuations for buying, selling, or leasing properties.
2. Valuing properties for tax assessments or legal disputes.
3. Valuing properties within real estate investment portfolios for performance measurement.
4. Assessing land value and development potential.
5. Determining replacement costs for property insurance purposes.
6. Valuing properties for corporate balance sheets and mergers/acquisitions.
7. Valuing properties subject to government acquisition.
8. Factoring sustainability into property valuations.
9. Complying with ethical standards during valuation and report writing.
10. Valuing niche properties such as heritage sites, hotels, airports or luxury resorts.

Today, software is configured to support use cases such as applying AI to automate processes or to find information. Use cases continue to emerge to improve the output and/or reduce turnaround times. To understand how these advances are possible, we begin by exploring the fundamentals of quantum computing and the technologies driving this transformation.

Demand for quantum technologies

The United Nations declared 2025 as the *International Year of Quantum Science and Technology*, and their potential demand is staggering: "It is estimated that the potential economic value for the trio of quantum computing, quantum sensing, and quantum communication and security in the leading industries could reach between $900 million and $2 trillion by 2035" (World Economic Forum 2025, p. 5). This demand is represented in the drivers for quantum technologies, like innovation opportunities to gain a competitive advantage or more secure systems that improve trust and privacy (World Economic Forum, 2025).

The demand for quantum technologies is forecasted to increase job growth; for example, in Australia, the quantum technology industry has the potential to directly employ 19,400 people, with its adoption expected to generate over 35,000 jobs across the economy by 2045 (Australian Government, n.d.). By 2045, Canada's quantum sector is projected to grow into a $139 billion CDN (content delivery network) industry, creating over 200,000 jobs and generating

$42 billion CDN in returns, potentially accounting for 3% of the country's GDP (Government of Canada 2022, p. 4). 126,100 quantum technology-related jobs may be created in the UK by 2045 (Law 2025). Similar forecasts are made for other countries and can be found online.

Quantum technologies promise capabilities to provide unrivaled innovation opportunities. Many countries, like Australia, forecast macroeconomic benefits by adopting quantum technologies: "Quantum could create sustainable, inclusive and well-paid jobs. By 2030, the Australian quantum sector could create around 8,700 jobs, rising to 19,400 by 2045. The growth of the sector would also support employment across the wider economy, with a total investment impact of over 35,000 jobs by 2045." (Australian Signals Directorate 2023). More countries are developing quantum technology strategies, ranging from Japan to Saudi Arabia. While quantum technologies have tremendous promise, they also present severe risks to most organisations, large and small.

The nightmare scenario for data security

Hackers also desire quantum computers since these technologies will give them access to an organisation's data (Table 13.1, Skulmoski et al., 2025). Using Shor's algorithm with a cryptographically relevant quantum computer, a hacker will be able to break most classical encryption (e.g., the "locks") that keep our systems and data secure. Currently, it would take a hacker around

Table 13.1 Data at risk from a quantum hacker

Organisational Data at Risk		
Customer information	*Bank account information*	*Loan information*
Income statements	Financial statements	Transaction details
Financial forecasts	Credit card statements	Payment card information
Employee information	Tax information	Wire transfer details
Asset valuation details	Formulas and recipes	Manufacturing procedures
Business strategy	Research and development	Computer source code
Pricing details	Technical data	Military strategy and tactics
Contracts and agreements	Intellectual property	State secrets
Minutes	Emails	Etc.
Personally Identifiable Information at Risk		
Full name	Social security number	Date of birth
Home address	Email address	Phone number
Driver's license number	Passport number	Banking information
Military identification	Photographs	Etc.

100–300 trillion years to hack today's popular cryptography (e.g., 2048-RSA encryption). A quantum hacker may be able to break into a system, steal and then decrypt the data in less than one day.

This data is at risk when at rest, in use and in transit, like in blockchain transactions. Therefore, this data will need improved protection with quantum-resistant cryptography (e.g., new locks).

Organisations should be migrating to quantum-resistant cryptography to protect their data. However, most organisations at the time of writing have not begun their quantum cybersecurity projects to migrate from quantum-vulnerable to quantum-resistant cryptography (new quantum locks to protect data and systems). Adoption studies are only now appearing. For example, an IDEMIA Group (2025) study about quantum technologies adoption found that around 40% of their organisations (n=35) had started their quantum journeys. A General Dynamics Information Technology (2024) study of American government adoption of post-quantum cryptography found that 50% of the agencies had started their post-quantum cryptographic migration journeys (n=200 cybersecurity subject matter experts). Thus, there is evidence of organisations adopting quantum technologies, including migrating to post-quantum cryptography.

Moody's, the risk management and insurance company, advises a programme of projects to implement post-quantum cryptography to protect the organisation's data. They predict these projects will be "long and costly" and could take organisations 10–15 years to protect against quantum threats (Moody's 2024). In the *Quantum Threat Timeline Report 2024*, Mosca wrote, "The progress in the last year has induced many people both within outside the quantum research community to realise that the quantum threat may be closer than they thought." (Mosca and Piani 2024, p. 5). Indeed, the White House directed its agencies and departments to begin their quantum journey: "The United States *must* prioritise the timely and equitable transition of cryptographic systems to quantum-resistant cryptography, with the goal of mitigating as much of the quantum risk as is feasible by 2035" (The White House, 2022, p. 2). Since the White House policy planners are *prioritising* the work to be completed in 13 years, it is a subtle affirmation about the complexity and duration of protecting the American government's prioritised data, and not even all the government's data!

The Paramount Milestone: Q-Day

Q-Day is the moment when a cryptographically relevant quantum computer becomes powerful enough to break widely used encryption algorithms, like RSA-2048, threatening global data security. This milestone could render current encryption obsolete, making the transition to quantum-safe cryptography essential to protect sensitive information. Mosca and Piani (2024) publish annual research results about the opinions of quantum technologies thought leaders (e.g., Peter Shor of MIT and Jay Gambetta of IBM), including range estimates for the arrival of a cryptographically relevant quantum computer. Each year, they report the respondents' estimates as to when a quantum computer can break RSA-2048 encryption in 24 hours. These quantum technology subject matter experts predict there is a 60% probability that a cryptographically relevant quantum computer will arrive around 2042 (Mosca and Piani 2024, Figure 26, p. 67). However, like the introduction of ChatGPT in 2022, the arrival of Q-Day may arrive sooner than anticipated and unexpectedly.

The dual potentialities of quantum technologies

There are dual potentialities of quantum technologies: they can bring transformative and innovative capabilities to healthcare, defense, finance, engineering, the built environment, and property and asset valuation. However, hackers can use quantum computing to break into and disrupt most systems and steal most data. Organisations should be developing dual strategies to plan, implement and optimise quantum technologies: innovate and protect. Valuers and others in the built environment can anticipate these technologies and learn how they work and how they might be used.

As property markets become increasingly influenced by numerous variables, quantum technologies can enable better forecasting and more accurate valuations, creating a competitive edge for investors and stakeholders. However, alongside these advancements come significant risks. The most pressing concern is the potential for quantum computers to undermine current encryption systems, posing a major threat to data security. Additionally, the prohibitive cost of research, development, and infrastructure, as well as the scarcity of specialised talent, create barriers to broader adoption. Balancing these challenges with the growing demand for innovation will be crucial as industries embrace the future of quantum technologies. Indeed, organisations may prioritise projects to protect their data from a quantum computer hack rather than implement technologies to improve valuation capabilities.

Quantum technologies 101

To simplify what Albert Einstein[1] referred to as "spooky action at a distance," project teams think of quantum technologies as hardware and software that they integrate and implement. They leave the quantum mechanics and engineering to physicists, computer scientists, electrical engineers, and mathematicians. In this book chapter, we investigate what quantum technologies can provide to the built environment and to the property and asset valuation profession.

Generally, there are three types of quantum technologies (hardware):

1 **Quantum sensors** offer unmatched accuracy and precision with applications such as forecasting earthquakes, improving mining safety, and enhancing

weather predictions. In the built environment, expect more precise subsurface quantum imaging for groundwater identification or an improved topographic survey of a pre-construction site dense with foliage. Researchers at the University of Birmingham have developed a quantum gravimeter that can create detailed maps of subterranean structures, such as pipes, ducts, and sinkholes, aiding efficient planning and reducing unforeseen obstacles during construction projects. This quantum sensor is being commercialised at the startup Delta-g (Quantum Technology Hub, 2024). Quantum computing and LiDAR may be combined to enhance sensitivity and accuracy in target detection and range estimation (Torromé and Barzanjeh 2024).

2 **Quantum communications**, or the "quantum internet," enable ultra-secure communications by using quantum mechanics to exchange encryption keys via light properties (photons), instead of mathematics. This quantum key distribution (QKD) technology will also enhance quantum sensing and computing, as it will be a robust communications backbone for other technologies. For example, drones will operate using the quantum key distribution network for ultra-secure operations. Quantum key distribution networks are beginning to be used, like the 46 km quantum key distribution network in Singapore that links two JP Morgan Chase data centers (JPMorgan Chase, 2024).

3 **Quantum computers** differ from classical computers in architecture and data processing (Figure 13.1, Skulmoski et al., 2025). While classical computers use bits (0 or 1), quantum computers use qubits, which can represent multiple states simultaneously (e.g., 65% zero, 35% one). This allows quantum computers to process data exponentially faster: simultaneous rather than consecutive calculations. As a result, quantum computing can process massive volumes of data required to unlock new innovations like developing new materials (e.g., transparent concrete). Quantum computers can be installed on the premises of an organisation (e.g., Cleveland Clinic hospital, USA) or through a quantum computing provider offering Quantum as a Service, where quantum computing resources (including technical support) are paid for by subscription. The Cleveland Clinic quantum computer from IBM is supported with $500 million USD in funding from the clinic and government programmes (Smith-Goodson 2023). Quantum computing subscriptions for Quantum as a Service (QaaS) range between $1000–$2000 USD per hour (Goldstein 2025), depending upon the processing time and support.

However, quantum computers used by hackers will be able to break our most common passwords (RSA-2048 encryption). There is a cybersecurity arms race to implement quantum-resistant cryptography (new locks) before hackers access a quantum computer that can break our current encryption .

(old locks)

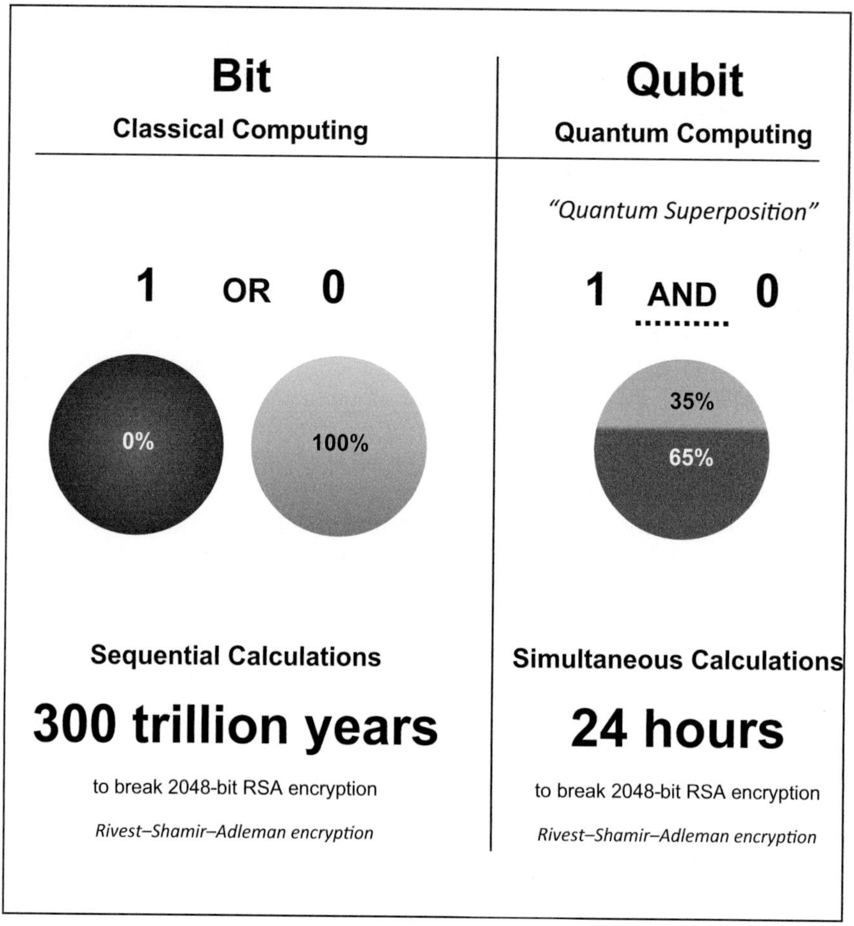

Figure 13.1 Classical vs. quantum computing.

The built environment will likely first see quantum sensors and quantum computers, followed by regional and then national quantum internet communication networks (quantum key distribution). Quantum communication networks and computers will create a technical ecosystem required for autonomous and flying vehicles, robots, and other equipment on the construction site. Valuators will use quantum computing to automate and optimise valuation processes and outputs.

Building quantum computers is challenging due to fragile quantum phenomena affected by decoherence, which leads to calculation errors. Fault-tolerant technologies aim to mitigate this issue. In 2024, Google's Willow chip achieved a milestone by reducing error rates exponentially while scaling qubits, marking significant progress towards quantum advantage (e.g., when a quantum computer can outperform calculations by classical computers).

With more research and development, quantum technologies are maturing, resulting in more use cases and quantum computing applications in general, and specifically for property and asset valuation. Many of these optimisation use cases will be based on artificial intelligence technologies tuned for quantum computing.

Adoption is underway in high-investment sectors such as healthcare and finance. Widespread industry use is within 10–20 years, contingent on reductions in cost, regulatory standardisation, and skills development. Traditional early adopters (e.g., finance, telecommunications, and energy industries) are poised to be the first to begin their quantum journeys (World Economic Forum, 2025, p. 5).

Quantum computing, artificial intelligence and valuation

There is a race to develop a quantum computer more effective than a classical computer (the *quantum advantage*). Once a sufficiently powerful quantum computer is available, it will be paired with software to direct the computer's performance. Some quantum-enabled software will be the next generation of artificial intelligence applications (quantum machine learning) to provide automated valuation models, real-time market analysis, predictive analytics and more.

Today and tomorrow, artificial intelligence with quantum computers will use big data as an input to solve problems and to optimise outcomes (e.g., minimise and maximise). Big data includes very large and complex data sets, including some that may appear to be unrelated; non-property data, analysed with quantum computing and AI technologies, may reveal market influences, enhancing valuation accuracy. A key point is that there is so much data to analyse, it would take classical computers millions to trillions of years to provide optimal solutions, whereas a quantum computer may complete those same calculations in an hour.

A partial list of big data may be seen in Figure 13.2 (Skulmoski et al., 2025), where future valuation professionals will use a quantum computer and quantum AI algorithms with big data to run their use cases.

This big data and more will be available, analysed with artificial intelligence and powered by a quantum computer to provide valuation practitioners with optimal solutions, valuations, risk assessments, market analyses, future cash flow projections, and so forth.

Four big AI questions

Artificial intelligence is a collection of technologies, including machine learning. Machine learning comprises systems that can perform tasks, then learn and adapt to improve performance, like predictions over time. Machine learning uses big data as an input to provide decision-making information

Figure 13.2 Big data (partial list).

(Figure 13.3, Skulmoski et al., 2025). AI and quantum computing will excel at providing valid and reliable answers to the big four AI questions:

1 Descriptive: what happened?
2 Diagnostic: what really happened?
3 Predictive: what will happen?
4 Prescriptive: what should I do?

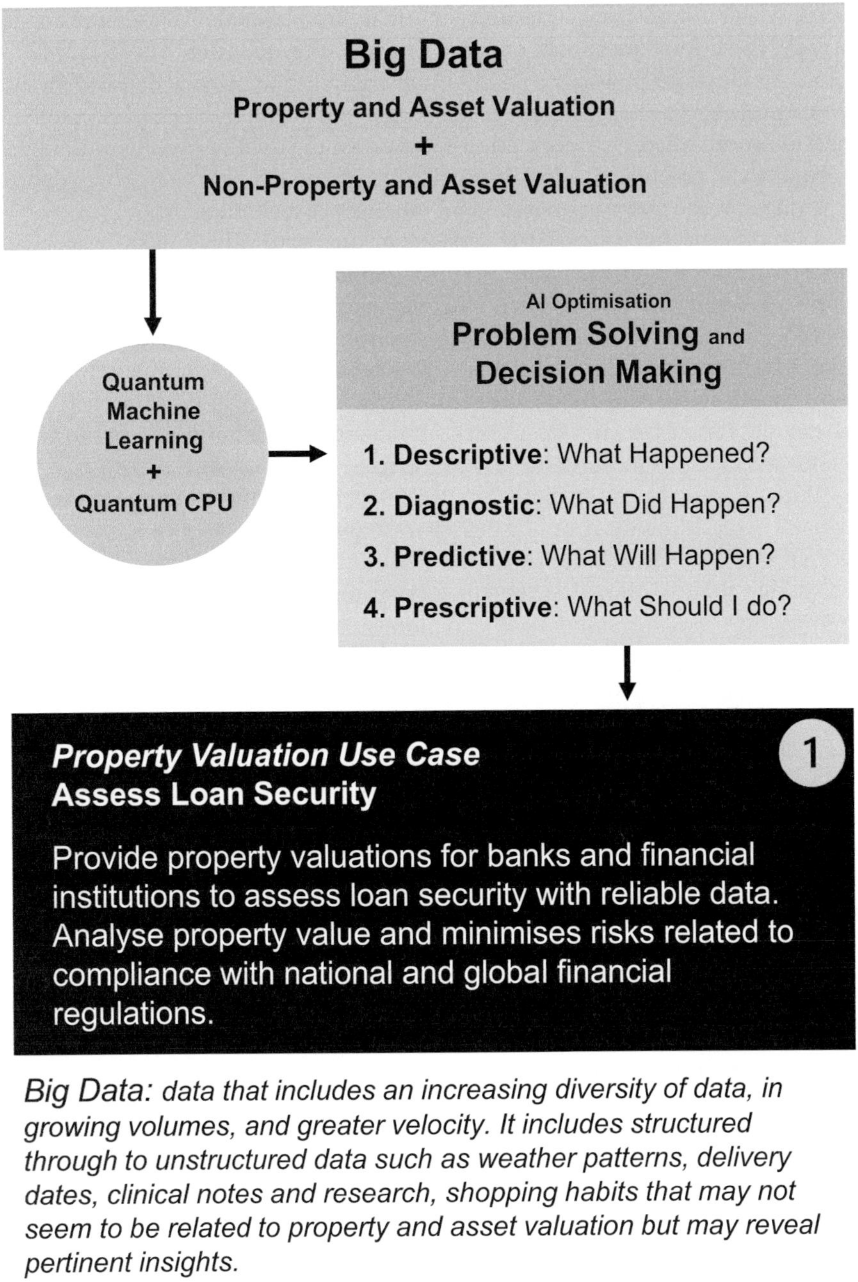

Figure 13.3 Assess loan security quantum use case.

Who hasn't asked these questions in their professional life? Therefore, to answer these four questions, big data is used with quantum artificial intelligence applications powered by a quantum computer to provide detailed analysis to property and asset valuers. All four types of analysis (e.g., predictive) provide comprehensive information to improve the quality of valuations (like accuracy and reliability), which is expected to improve over time due to machine learning and the maturing of artificial intelligence technologies.

Like other technologies, early versions of quantum valuation software may have defects or shortcomings that are addressed with successive quality improvement cycles and through machine learning. Think of the first version of *MS Word for DOS* in 1983 or the first automobiles that were incredibly different from today's versions of those early technologies. Quantum computing with quantum AI will be no different; early versions will be improved, and after a decade or two, the quantum technologies and computing used in valuations will become more stable and mature.

With apologies to technology architects for our over-simplification (Figure 13.3, Skulmoski et al., 2025), a quantum computer (hardware) will power artificial intelligence software (quantum machine learning) that will access big data to provide problem-solving, modelling and decision-making information used to assess assets for loan security. This quantum machine learning architecture includes data collection, valuation model development and training, and then use by evaluators. AI model training (optimisation) is continuous and powered with a quantum computer, resulting in improved outcomes (e.g., predictions, valuations and reports) over time.

Asset and property valuers do not work in a vacuum; instead, sometimes they collaborate with other professionals to achieve common goals like developing a legal contract for a merger and acquisition. For example, an asset may need to be valued, and a contract developed; however, the process is complex when international contracting partners cover multiple legal authorities (Figure 13.4, Skulmoski et al., 2025). That requires asset valuation as an input; the output will be an optimised contract. Big data in Figures 13.3 and 13.4 can be used with quantum AI and quantum computing to optimise the valuation and the merger and acquisition contract to minimise risks.

When applying quantum technologies to property and asset valuation, look to AI use cases for classical computing that can be tweaked with quantum algorithms. The goals and use cases are much the same whether AI for classical computing or AI for quantum is used: the goal is to develop valid and reliable valuations with a minimum amount of bias. With quantum AI (quantum machine learning) and extensive data integration, valuators will receive real-time updates for improved data accuracy and predictions. No doubt, quantum technologies will lead to other innovations and use cases of interest to the valuation community.

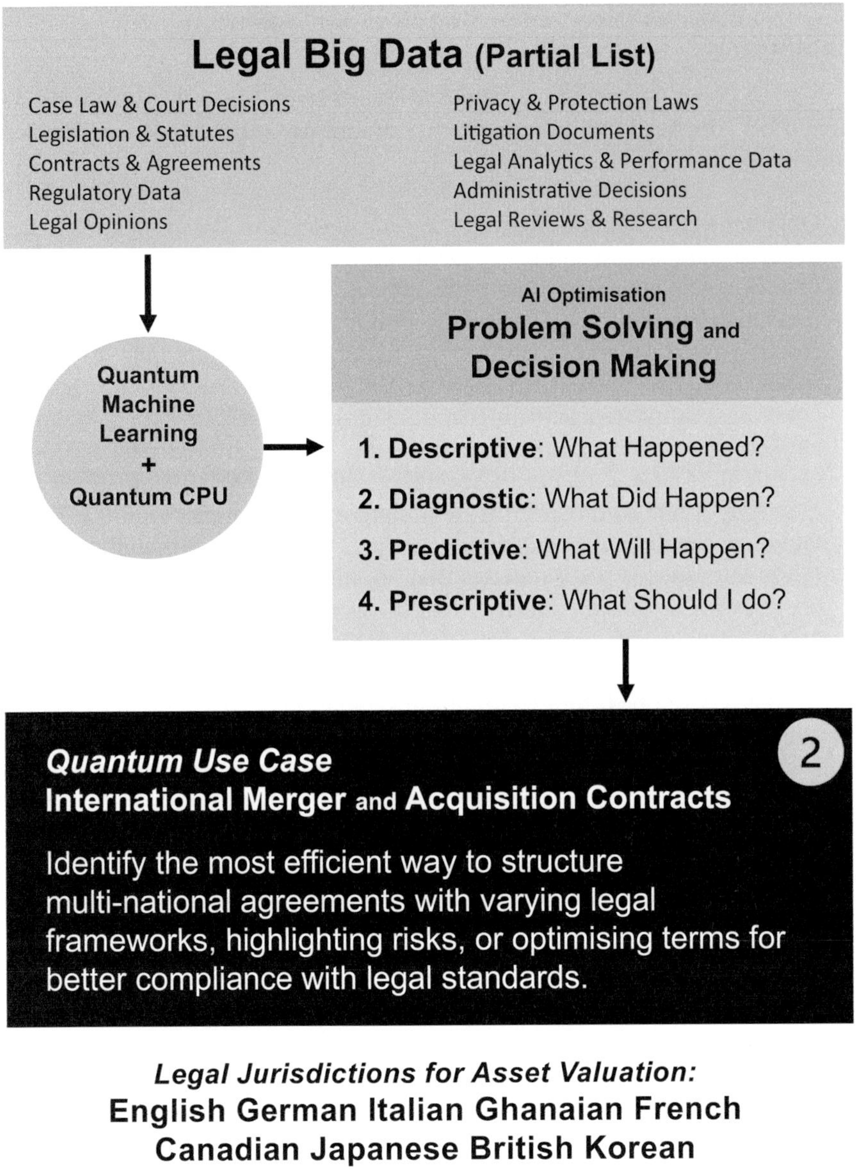

Figure 13.4 Contract development with quantum technologies.

Beyond optimisation: Emerging uses for quantum computing

Quantum computers are emerging, and their capabilities will expand. There are other applications of quantum computing other than strictly for optimisation:

1. **Quantum search**: Quantum computing and algorithms can speed up searches of large data sets to improve search results (e.g., improved accuracy and quality). Applying quantum algorithms to search for the best solution (maximise or minimise), including combinatorial optimisation (the best solution from a finite number of permutations/combinations), and algorithms to speed up searches (Grover's algorithm to search unstructured databases). Improved searching through large numbers of data sets can be applied to quantum optimisation use cases, like in multinational contracts (Figure 13.4); here, quantum searching finds the right information, and quantum optimisation finds the minimums and maximums.
2. **Quantum simulation**: Simulation techniques to mimic and model quantum molecular systems and processes that are intractable for classical computers, allowing new insights such as drug discovery and new materials. Property and asset valuators are unlikely to *create* new materials but will need to value assets that include new materials like self-repairing polymer roofing materials developed through quantum simulations.
3. **Quantum factorisation**: The process of using quantum computers and algorithms (e.g., Shor's algorithm) to factor exceptionally large integers into their prime components is problematic for classical computers and the basis for classical encryption (e.g., RSA-2048). Hackers will use quantum factorisation to break classical encryption, like which keeps blockchain transactions secure. Therefore, organisations should migrate to post-quantum cryptography to keep their data and systems secure. Migrating to quantum-safe systems is likely to take priority over implementing quantum and AI valuation systems.

Barriers to adoption

Despite their innovative opportunities, there are significant barriers to quantum technologies adoption, including:

1. **Lagging regulatory frameworks**: Organisations may delay their quantum technologies investment and adoption until regulatory environments mature. However, organisations can develop quantum technologies awareness and knowledge while regulations develop.
2. **Quantum obliviousness**: Some organisations may not be aware of quantum advancements and their potential impact, necessitating awareness initiatives to address this gap.

3 **Lack of quantum competencies**: A global shortage of quantum-skilled individuals may delay projects, but competence can be developed through other means, like training, partnerships, and community involvement.
4 **Immature quantum industry**: The early growth stage of quantum technologies, with its inherent risks and uncertainties, may lead organisations to delay adoption until the industry matures and standardisation advances.
5 **Emerging data privacy and security environment**: As the quantum industry matures, organisations can rely on resources like National Institute of Standards and Technology's (NIST) Privacy Framework to guide data privacy and security practices.
6 **Weak cross-industry engagement**: Collaboration opportunities in the emerging quantum sector may be limited due to competition, but strategic collaborations can yield mutually beneficial results.
7 **Quantum hesitancy**: Indecision or delay in adopting quantum technologies due to uncertainty can be addressed with awareness initiatives, particularly around risks of a cryptographically relevant quantum computer.
8 **Financial**: Limited budgets may hinder full quantum adoption, but lower-cost activities like experimenting with quantum algorithms can provide valuable insights before larger investments are made.
9 **Near-term focus**: Some organisations prioritise near-term projects like AI and zero-trust cybersecurity due to limited resources and higher priorities, delaying quantum initiatives and potentially requiring compressed schedules to catch up.

Indeed, most organisations face multiple barriers, like focusing on near-term projects and lacking quantum technology competencies. Therefore, many organisations have yet to start their quantum projects due to adoption barriers like quantum hesitancy, requiring project champions to overcome these barriers to embrace quantum technologies.

Managing quantum technologies risks

Any risks related to planning, implementing and operating these new technologies can be managed like other technology lifecycle risks (Figure 13.5, Skulmoski et al., 2025). For example, begin by updating and aligning the organisation's key strategies—business, technology and cybersecurity strategies—to adopt these technologies. Develop a business case (e.g., quantum sensing for groundwater discovery) and secure approval to proceed to plan and implement the technologies outlined in the business case. A technology project follows that aligns with best practices (e.g., proactive risk management as outlined in ISO 31000 Risk Management). The project deliverables (quantum sensing) are handed over to the project sponsor and used according to the business case to generate value for the organisation and customers. Once the quantum sensors are used, the technology department follows technology

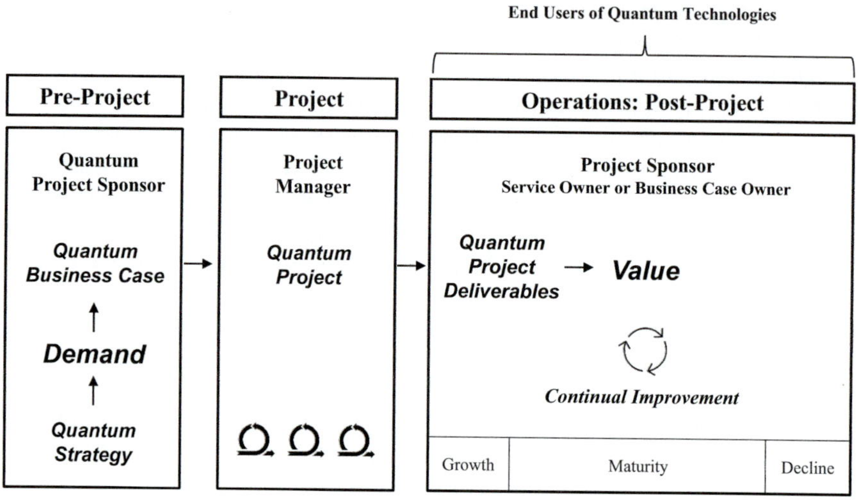

Figure 13.5 Quantum technologies lifecycle.

service management practices (e.g., incident reporting to the service desk). These practices are found in the Information Technology Infrastructure Library (ITIL) to manage these technologies while reducing risks and achieving the intended quality levels.

Therefore, manage quantum technology risks (pre-project, project and post-project) by applying, tailoring and combining best practices outlined in globally accepted standards and frameworks.

Conclusion

Property and asset valuation professionals have long adapted to new technologies, from paper-based ledgers to spreadsheet software and, more recently, to cloud-based AI-enabled platforms. As we approach a new technological inflection point, quantum technologies promise to be even more transformative. These technologies—particularly quantum computing, quantum sensing, and quantum communication—are poised to reshape the valuation profession by enabling more accurate, timely, and reliable decision-making, especially when applied to complex and high-dimensional datasets.

The integration of quantum computing into valuation practice offers an exponential increase in computational speed and analytical depth. This leap empowers valuers to forecast markets in real time, assess complex and interconnected risks, and embed sustainability metrics into their assessments. More significantly, quantum machine learning (QML) enables the development of predictive and prescriptive valuation models that can self-optimise over time, yielding enhanced insights with reduced bias. As with the introduction of artificial intelligence, those who embrace quantum technologies early will likely secure significant competitive advantages and deliver greater value to their clients and stakeholders.

However, the very capabilities that make quantum computing powerful also introduce new risks, particularly in cybersecurity. The anticipated arrival of a cryptographically relevant quantum computer—often referred to as Q-Day—poses a substantial threat to existing encryption protocols, including those underpinning sensitive valuation data, contracts, and financial transactions. Organisations must begin migrating to quantum-resistant cryptography now to mitigate these looming risks. This dual reality—unparalleled innovation coupled with unprecedented vulnerability—requires a careful and proactive approach to technology adoption.

Successfully navigating this new frontier involves more than just acquiring the latest tools. Valuers must develop quantum literacy and incorporate best-practice frameworks from strategy, project management, and cybersecurity. The lifecycle management of quantum technologies—from strategic alignment and business case development to implementation and service management—ensures not only value realisation but also the mitigation of operational and security risks. Leveraging standards such as ISO 31000 for risk management and ITIL for service delivery provides a structured pathway for responsible innovation.

Despite the promise, several barriers may delay adoption. These include lagging regulatory frameworks, talent shortages, quantum hesitancy, and financial constraints. Many organisations remain focused on near-term priorities, such as artificial intelligence and zero-trust cybersecurity, leaving quantum initiatives under-resourced or postponed. Addressing these barriers requires project champions, strategic vision, and cross-industry collaboration.

Nevertheless, the use cases for quantum technologies in valuation are emerging and compelling. Table. 13.2 (Skulmoski et al., 2025) outlines current and near-future applications where quantum computing and quantum communications can deliver tangible improvements in the valuation lifecycle.

Table 13.2 Quantum technology valuation use cases

Valuation Use Case	*Quantum Technology*
Property valuation	Quantum computer—optimising and searching
Market forecasting	Quantum computer—optimising and searching
Risk assessment	Quantum computer—optimising and searching
Sustainability analysis	Quantum computer—optimising and searching
Contract structuring	Quantum computer—optimising and searching
Report writing	Quantum computer—optimising and searching
Secure communications	Quantum key distribution

Valuers, like many professionals in the built environment, do not work in isolation. They are part of an interdisciplinary network that includes legal, financial, technological, and regulatory actors. Collaboration within this ecosystem will be essential to both leverage the potential of quantum technologies and manage their associated risks. As quantum computing becomes more accessible through models such as Quantum as a Service (QaaS), early experimentation with quantum algorithms and participation in communities of practice can help professionals build experience and reduce future transition costs.

In sum, quantum technologies offer dual potentialities: the ability to radically innovate and the imperative to secure and protect. Property and asset valuers should begin their quantum journey now—not simply to avoid obsolescence, but to shape the future of their profession. By combining a readiness to innovate with a disciplined approach to risk, valuers will be well-positioned to deliver trusted, high-quality services in an increasingly complex and data-driven world.

There are many opportunities to learn more about quantum technologies, including:

- Property and asset valuers can find many resources online to develop their knowledge of quantum technologies as they begin their quantum journeys.
- Follow quantum technology companies and thought leaders on social media and regularly read something about quantum technologies.
- Learn more about the cybersecurity threats from someone misusing quantum technologies.
- Investigate how blockchain technologies will need to be updated with post-quantum cryptographic migrations.
- Look for other quantum champions to collaborate with in your value chain.
- Join or start a quantum technologies community of practice.

Questions

- How might quantum computing change the way I assess complex property portfolios or conduct market forecasting compared to traditional methods?
- Which specific quantum use cases for process optimisation seem most relevant to my valuation work, and why?
- What are the implications of quantum-enabled cybersecurity threats for the confidentiality and integrity of the valuation data I handle?

Readings:

- Greg Skulmoski (2025), Accelerated Quantum Technologies Change Management, Business Expert Press, New York.
- Greg Skulmoski and Ashkan Memari (2025). Quantum Cybersecurity Program Management (In Publication), Business Expert Press, New York.

References: Quantum learning resources

There are online resources to learn about quantum technologies. Follow groups and thought leaders on LinkedIn and other social media platforms. Use generative AI to find out more about quantum technologies in your discipline. The following curated resources can also provide a quick start for your quantum journey.

Glossary

Our glossary is developed for practitioners. Purists may find we may have tailored our definitions a bit too much. Still, we take some liberty to tailor and combine to create an applied glossary for readers who are new to quantum technologies. The reader is encouraged to go online to find or generate additional information.

Algorithm a set of instructions (e.g., a suite of quantum circuits) to solve a problem or perform calculations, and can be thought of as software used by a quantum computer (hardware).

Classical cryptography cryptography that is secure against classical computing but not quantum computers (see also *Post-quantum cryptography*).

Combine bring together tools and processes from different standards and frameworks.

Competence the knowledge, skills and abilities that are effectively applied by the people on the project teams (see *Capability*).

Cryptographically relevant quantum computer quantum technologies with sufficient capabilities to break public-key systems (e.g., asymmetric cryptography) used to protect systems with classical cryptography. Mosca's "Z" parameter represents the arrival of a cryptographically relevant quantum computer.

Cryptography the study of ensuring the safe transfer and storage of information in an adversarial environment.

Cybersecurity protecting the organisation's information through risk management practices and processes (e.g., prevent, detect and respond to cybersecurity incidents).

Demand technology users request existing or new digital products or services, and it is detailed in ITIL's Demand Management practice.

Dual potentialities of technology technology is inherently neutral; however, it (quantum computing) can be used for good (curing diseases) or bad purposes (stealing and decrypting).

Early adopters individuals and organisations who use new technologies before others (see *Late adopters*).

Encryption the process of scrambling data so only the intended parties can unscramble it.

Encryptogeddon a point in time when the collective negative impacts of threat actors using quantum technologies like quantum computing to break encryption with brute force techniques outweigh the benefits of quantum technologies.

Frameworks represent best practices but usually exist in the absence of well-defined and globally accepted standards. Frameworks are less prescriptive and more flexible than standards.

ISO standards guides embodying international best practices agreed upon by experts. They often have a certification pathway indicating that they meet the required levels of quality to safely deliver the service or quality for which they are certified.

ITIL the Information Technology Infrastructure Library is a service management framework of practices that outline how to plan, deliver, operate, and optimise digital services (see *COBIT*, an alternative framework).

Late adopters individuals and organisations who use new technologies after others (e.g., after early adopters have stabilised the technology) and are sometimes referred to as laggard adopters (see *Early Adopters*).

Operations the phase after project deliverables (e.g., a quantum computer that simulates protein structure predictions in human health and disease research) are provided to the project sponsor, who manages the deliverables according to the quantum business case and to the benefit of the organisation.

Post-quantum cryptography cryptographic algorithms that are presumed to be secure from both quantum and classical computer attacks (also known as quantum-safe cryptography and quantum-resistant cryptography; see also *Classical cryptography*).

Q-Day arrival of a cryptographically relevant quantum computer that can break the 2048-bit RSA key widely used to keep data and systems secure and requires a transition to quantum-resistant cryptographic systems.

Quantum advantage a point in time where a quantum computer can solve a problem more efficiently and effectively (e.g., more accurately) than a classical computer, irrespective of the solution's utility. Quantum advantage is the preferred term, as quantum "supremacy" has negative connotations and is related to racial supremacy for some people.

Quantum business case a high-level or strategic description of the quantum product or service, product, or service manager (e.g., service owner), contribution to strategy, benefits, timelines, costs, return on investment and risks and issues.

Quantum business use case a detailed or tactical description of how quantum technologies are applied to a specific business problem or opportunity. A quantum business use case can be part of the quantum business case (e.g., contribution to strategy).

Quantum cryptography applying quantum properties to create cryptographic protocols like quantum key distribution.

Quantum key distribution quantum key distribution (QKD) encrypts and decrypts data with the principles of quantum physics (e.g., superposition and entanglement); however, post-quantum cryptography is generally recommended over QKD.

Quantum readiness the ability to manage threats related to quantum technologies in the domains of people, process and technology, including implementing post-quantum cryptography and developing and maintaining cryptographic agility. Quantum readiness assumes a minimum viable cybersecurity foundation is maintained and continually optimised for both quantum and classical technologies.

Quantum-safe cryptography the study of cryptographic algorithms that are thought to be secure from both quantum and classical computer attacks (also known as post-quantum cryptography and quantum-resistant cryptography).

Quantum technologies technologies that apply the principles of quantum physics, like quantum computers or quantum sensors.

Quantum vulnerable a system susceptible to a quantum attack (e.g., the digital ecosystem may have insufficient post-quantum cryptography).

Qubit the smallest unit of information in quantum computing, like a bit in classical computing, except a qubit can be in both states of 0 and 1 simultaneously (superposition property of quantum mechanics).

RSA cryptography Rivest–Shamir–Adleman (1977) is a public-key encryption algorithm for secure data transmission to encrypt data using two different but linked keys.

Simulation using quantum technologies, including quantum computers, to model and simulate physical and chemical systems at the quantum level.

Steal now, decrypt later to steal data and wait until quantum tools are available to decrypt the data (also known as record now, exploit later).

Tailor to bring in and adapt standards, frameworks, techniques, tools and processes suitable for one's project. In *Quantum Cybersecurity*, we consider tailoring to adapt within the framework, standard, or method and combining to bring in techniques, tools and processes from other frameworks, standards, or methods.

Value technology users' perception of the benefits of digital products or services, usefulness and importance that are provided by the organisation, and is central to ITIL's Service Value System and Service Value Chain.

Note

1 Albert Einstein wrote to Max Born in 1947 about quantum mechanics and entanglement with the translation from German is "physics should represent a reality in time and space, free from spooky action at a distance" [spukhafte fernwirkung]. In *The Born-Einstein letters: correspondence between Albert Einstein and Max and Hedwig Born from 1916–1955, with commentaries by Max Born*. Macmillan. 1971. p. 158.

References and suggested readings

Australian Government. (n.d.) "Australia's Quantum Opportunity," https://www.industry.gov.au/publications/national-quantum-strategy/australias-quantum-opportunity#:~:text=The%20quantum%20technology%20industry%20could,across%20the%20economy%20by%202045.&text=This%20graph%20shows%20the%20projected,the%20Australian%20economy%20by%202045

Australian Signals Directorate 2023. Planning for Post-Quantum Cryptography, https://www.cyber.gov.au/resources-business-and-government/governance-and-user-education/governance/planning-post-quantum-cryptography

General Dynamics Information Technology 2024. "Quantum Waves: Agency Guide to Post-Quantum Cryptography," https://www.gdit.com/perspectives/quantum-waves/pdf/

Goldstein, Phil 2025. "How Will QaaS Technology Become Available To The Masses? *BizTech* 13 January 2025, https://biztechmagazine.com/article/2025/01/qaas-technology-becoming-available-perfcon

Government of Canada 2022. "Canada's National Quantum Strategy," https://ised-isde.canada.ca/site/national-quantum-strategy/sites/default/files/attachments/2022/NQS-SQN-eng.pdf

IDEMIA Group 2025. "Key obstacles to Post-Quantum Cryptography (PQC) Adoption," 26 February 2025, https://www.idemia.com/insights/key-obstacles-post-quantum-cryptography-pqc-adoption?utm_source=chatgpt.com

JPMorgan Chase May 8, 2024. "JPMorgan Chase Establishes Quantum-secured Crypto-agileNetwork,"https://www.jpmorgan.com/technology/news/firm-establishes-quantum-secured-crypto-agile-network

Law, Marcus 2025. How Quantum Computing Could Add £212bn to UK Economy by 2045, *Technology Magazine*, https://technologymagazine.com/articles/how-quantum-computing-could-add-212bn-to-uk-economy-by-2045

Moody's 2024. "Transition to Post-Quantum Data Encryption will be Long and Costly," https://www.moodys.com/reports?topic=topic_f180dedc-7b16-11e9-9d45-1f411e81c7f7

Mosca, Michele, and Piani, Marco 2024. Quantum Threat Timeline Report 2024, Global Risk Institute, https://globalriskinstitute.org/publication/2024-quantum-threat-timeline-report/

Quantum Technology Hub 2024. "Reflecting On a Decade of Quantum Sensors and Timing,"https://www.quantumsensors.org/wp-content/uploads/2024/11/QT-Hub-10th-anniversary-report-optimised.pdf

Skulmoski, Greg 2025. *Accelerated Quantum Technologies Change Management*, Business Expert Press, New York.

Skulmoski, Greg and Ashkan Memari 2025. "Get Ready for the Next Generation of AI Quantum Technologies," Marx, E.W. (Ed.). *Voices of Innovation: Fulfilling the Promise of Information Technology in Healthcare* (3rd ed.), Boca Raton, FL: Productivity Press. https://doi.org/10.4324/9781032715155

Smith-Goodson 2023. "Cleveland Clinic And IBM Launch World's First Quantum Computer Dedicated To Healthcare Research And Biomedical Discoveries," *Forbes*, 21 March 2023, https://www.forbes.com/sites/moorinsights/2023/03/21/cleveland-clinic-and-ibm-launch-worlds-first-quantum-computer-dedicated-to-healthcare-research-and-biomedical-discoveries/

The White House 2022. National Security Memorandum on Promoting United States Leadership in Quantum Computing While Mitigating Risks to Vulnerable Cryptographic Systems, May 04, 2022, https://www.whitehouse.gov/briefing-room/statements-releases/2022/05/04/national-security-memorandum-on-promoting-united-states-leadership-in-quantum-computing-while-mitigating-risks-to-vulnerable-cryptographic-systems/

Torromé, Ricardo Gallego, and Shabir Barzanjeh 2024. "Advances in Quantum Radar and Quantum LiDAR." *Progress in Quantum Electronics* 93: 100497-. https://doi.org/10.1016/j.pquantelec.2023.100497

World Economic Forum 2025. Embracing the Quantum Economy: A Pathway for Business Leaders, https://reports.weforum.org/docs/WEF_Embracing_the_Quantum_Economy_2024.pdf

Chapter 14

Valuation beyond the divide

Rethinking professional practice and standards in an era of digital transformation

Johari Hussein Nassor Amar

Introduction

The real estate valuation profession stands at a pivotal juncture amid rapid advances in artificial intelligence (AI), LiDAR scanning, drones and other similar technologies. These innovations are not merely incremental tools – they are fundamentally reshaping how valuers collect data, analyse information and render professional judgements. The stakes are high given that real estate assets constitute over half of global wealth, meaning that how valuers harness (or resist) new technologies can ripple through mortgage lending, investment decisions and financial stability. A cautionary example is real estate company Zillow's widely publicised failure in algorithmic home-buying, where an over-reliance on an AI pricing model led to losses of around US$880 million (Amar, Chapter 1; Crawford & Brogran, Chapter 2). This high-profile setback highlights the tension between computational capability and professional judgement, underscoring that while data-driven models offer unprecedented opportunities to enhance efficiency and analytical depth, their misuse or unchecked optimism can undermine valuation reliability and market trust. Ultimately, how valuers navigate this technological turn will shape the transparency and stability of property markets, making it crucial that these tools are integrated prudently and responsibly.

Rather than viewing AI as a magic solution or an existential threat, Wu and Wong (Chapter 4) argue that the emerging consensus is that the future of valuation lies in a balanced partnership between human valuers and intelligent machines. Advanced digital tools are most powerful when guided by seasoned professional expertise and ethical oversight. Each technology extends the valuer's reach in distinct ways: AI algorithms can rapidly sift through large datasets to identify market patterns and generate preliminary value estimates far faster than any individual can; LiDAR systems capture precise three-dimensional measurements of buildings and sites, yielding spatial data far beyond what traditional hand measurements provide; and drones can conduct aerial inspections of properties or infrastructure, safely producing comprehensive visual records of conditions that might be dangerous or impractical for a person to

DOI: 10.1201/9781003539254-18

assess directly. These capabilities dramatically augment the scope and speed of analysis. However, they do not diminish the need for the valuer's critical judgement; if anything, they elevate the role of a valuer as the interpreter and validator of technologically derived insights.

Throughout the case studies (refer to Section 3), it is clear that these tools redefine rather than replace the valuer's role. AI may automate calculations or standardise certain analytical steps, but a valuer is still needed to interpret results, ensure they make sense in context and make the final determination of value. LiDAR can provide unprecedented measurement accuracy, yet human expertise must decide how to incorporate that data into a valuation model and account for any anomalies or uncertainties. Drones might monitor property conditions continuously, but professional oversight is required to verify findings and navigate any legal, privacy or safety issues arising from automated inspections. In short, Crawford and Brogan (Chapter 2) state that technology works best as an assistant to the valuer through handling laborious or complex tasks, with the practitioner remaining as the accountable decision-maker who provides context, qualitative judgement and adherence to ethical standards. This augmentation of practice is an evolution of professional expertise, not a dilution of it: it allows valuers to leverage new capabilities while upholding the rigour, impartiality and caution that define professionalism in valuation.

The digital transformation of valuation is playing out on multiple levels: individual, organisational, industry-wide and regulatory. At the individual level, Fonteyn (Chapter 13) observes that valuers are adapting to new skill requirements, from data science literacy to operating drones, and are shifting their professional identity from being sole experts to becoming collaborative analysts working alongside algorithms. At the organisational level, Wallace, Wallace, Moorhead and Logue (Chapter 6) note that firms are integrating these technologies into workflows and service models, for example, by adopting centralised AI platforms or investing in in-house drone and LiDAR capabilities while also managing training, costs and quality control. At the industry level, Khanjanasthiti, Gardner and Taklif (Chapter 3) adoption patterns vary; some market sectors and regions enthusiastically embrace mass appraisal models and remote sensing, whereas others remain cautious due to cultural norms or risk concerns. Finally, at the regulatory level, Cradduck (Chapter 11) notes that professional bodies and governments are developing comprehensive guidelines to govern AI-assisted practices, from ensuring auditability of automated reports to establishing clear accountability frameworks for human-AI collaboration. The successful digital transformation of valuation demands alignment across all these scales. A valuer can only fully leverage AI and other tools if, for instance, their firm provides the necessary technology and training, if clients and regulators accept algorithm-assisted reports, and if clear professional guidelines exist to maintain accountability. In this sense, integrating technology is as much a management and governance challenge as it is a technical one.

Overall, the message of this volume is optimistic. If implemented thoughtfully, digital augmentation can indeed transform valuation practice for the better; i.e., improving efficiency, consistency and insight while preserving the accuracy, fairness and public trust that underpin the profession. The sections that follow synthesise the book's key thematic findings, outline practical implementation pathways and barriers, discuss implications for professional development and quality assurance and propose a forward-looking research agenda to guide the valuation profession's journey into the digital age.

Key themes of digital transformation in valuation

The chapters collectively reveal several major themes in how technology is reshaping valuation practice. These range from the practical integration of tools into workflows to the evolving skill set of valuers, the paramount importance of data integrity and the global equity implications of innovation.

Technology integration in practice

A unifying theme across the case studies is the increasing integration of AI, sensor platforms and drones into core valuation workflows. Traditional valuation practice relied heavily on in-person site inspections, manual data collection and spreadsheet-based analysis. Now, digital tools are becoming embedded at each step. For example, Fonteyn (Chapter 12) describes how valuers use big data analytics and machine learning (ML) models to enhance the analysis of comparable sales, market trends and property attributes. Automated valuation models (AVMs) driven by AI can instantly process large databases of sales to produce initial value estimates for residential properties. For data collection, LiDAR scanners and drones are streamlining how physical property information is gathered, where a drone flyover can capture high-resolution imagery or generate a 3D model of a site, while a LiDAR unit produces detailed measurements of buildings and terrain, essentially creating a 'digital twin' of the property that can be examined remotely. Tasks in the valuation practice that once required physical inspections and hours of effort can now be completed faster, remotely and with digital precision. This represents a qualitative shift in practice, expanding what a single valuer can cover and the level of detail achievable within typical timeframes.

However, the evidence also shows that technological adoption is uneven and context-dependent. Identical tools may be embraced in one jurisdiction or market sector and resisted in another due to institutional and cultural factors. Chapter 3 provides a comprehensive international comparison, which found that the regulatory environment and data infrastructure strongly influence integration rates; for instance, some countries actively encourage the use of AVMs for mortgage lending or permit drones for mass appraisal work, whereas others restrict these tools because of privacy laws, liability concerns or lack of trust in automated systems. In certain jurisdictions, Khanjanasthiti, Taklif and Gardner (Chapter 7) drones are prohibited over residential areas but freely used for agricultural or industrial valuations where airspace is less contentious. Likewise, LiDAR scanning is becoming standard for complex commercial or mining properties that demand precise volumetric data, but it remains uncommon for routine house valuations in markets where cost or familiarity is an issue. In short, while advanced valuation technology is globally available, its uptake is locally moderated by factors such as legal frameworks, organisational readiness and client expectations.

The chapters in this volume reinforce that successful innovation requires more than just technology availability – an enabling environment of clear regulations, data-sharing mechanisms and proven use cases is needed to build stakeholder confidence. Over time, as best practices spread and standards evolve, these integration gaps may narrow. At present, though, the landscape is a mosaic because some firms and regions operate at the cutting edge of AI and remote sensing, and others proceed cautiously or not at all, depending on external drivers and internal preparedness.

Transformation of professional roles and skills

Rather than making human valuers obsolete, the technological wave is reshaping the competencies and activities that valuation work involves. Across the chapters, it is reiterated that valuation ultimately remains a human judgement; for example, Ge and Raptis (Chapter 5) reinforce that without the involvement of a qualified valuer, no AI-driven process can truly constitute a valuation, emphasising that human expertise remains the cornerstone of professional practice regardless of technological sophistication. What is changing is the pathway to that final judgement. The concept of augmented professionalism has been used to describe how human expertise is extended by machines rather than replaced. Wu and Wong (Chapter 4) argue that this shift towards augmented professionalism requires valuers to develop hybrid skills combining traditional expertise with data science literacy to effectively orchestrate human-AI collaboration. In practice, valuers in technology-rich environments are acting more as orchestrators and interpreters of analytical outputs instead of manually performing every calculation or measurement. For example, Khanjanasthiti, Taklif and Gardner (Chapter 7) highlight that a valuer might use an AI model to preprocess market data and suggest a preliminary value range, deploy a drone to conduct part of the property inspection, and utilise GIS and LiDAR data to quantify land or building dimensions, all feeding into their assessment. The valuer then focuses on integrating these inputs, scrutinising them for reasonableness and applying local market insight and professional scepticism to arrive at the final opinion of value.

Several chapters – for example, Wolf and Amar (Chapter 8), Boyd (Chapter 9) and Cradduck (Chapter 11) – predict that mundane or standardised valuation tasks will become highly automated, potentially reducing the number of valuers needed for routine appraisals, while elevating the importance of human input in complex and non-standard cases. In the near term, straightforward residential assessments might be performed mostly by algorithms under a valuer's supervision, allowing practitioners to devote more attention to specialised work such as litigation support, unique properties or the reconciliation of conflicting indicators in borderline cases. This points towards a bifurcation of practice: simple valuations handled largely by AI (with oversight), versus complex valuations where human expertise is even more critical. Several chapters predict that valuers will increasingly focus on heterogeneous, high-risk or judgement-intensive assignments and on providing the ethical, contextual and interpersonal elements of service that machines cannot (Chapters 4, 6, 10, 11 and 12). This implies the necessity of the broadening of professional skill sets, where valuers are increasingly working in multidisciplinary teams; for instance, collaborating with data scientists or software specialists when developing an in-house AVM or analytics tool (Ge & Raptis, Chapter 5). As a result, the professional identity of valuers is expanding from pure real estate experts to hybrid analyst-technologists. Education and continuous professional development programmes are already being reoriented to produce graduates comfortable with statistical models, coding basics, remote sensing techniques and data ethics. In summary, the profession is not being replaced but is being redefined. Automation is removing some traditional tasks, but, as reflected by Fonteyn (Chapter 12), it is simultaneously amplifying the importance of higher-order analytical, interpretive and advisory skills that only experienced valuers can provide, thereby solidifying the valuer's role as the crucial human arbiter in an AI-supported process.

Data quality, bias and model transparency

A critical theme running through the volume is the paramount importance of data quality and the closely related issues of bias and transparency in valuation models. Advanced analytical tools are only as reliable as the data and algorithms behind them. If an AI-driven valuation model is trained on biased or unrepresentative data, it can produce systematically skewed valuations; for example, as Wolf and Amar (Chapter 8) note, it might consistently undervalue homes in a particular minority neighbourhood because the training data reflects historical bias, posing serious risks to fairness and accuracy. Several contributors, for example, Ge and Raptis (Chapter 5), caution that without proper checks, automated valuations could inadvertently perpetuate social biases or produce errors that a human valuer would catch. Crawford and Brogan (Chapter 2) emphasise that AI hallucination, where systems generate plausible but inaccurate outputs, represents a fundamental limitation requiring valuers to maintain professional scepticism and validate all algorithmic results. Regulators are beginning to respond: Cradduck (Chapter 11) notes that some authorities have started issuing guidelines to improve the transparency and explainability of AVMs and AI tools used in valuation, aiming to ensure these systems can be audited and held to professional standards.

The case studies also show that practising valuers have to become vigilant about the outputs of black box models. It is not enough to accept an algorithm's number at face value; the column emphasises that professionals need to interrogate how that result was derived, cross-check it against market evidence and be alert to anomalies produced by AI responses (e.g. false or nonsensical output). The push for algorithmic transparency is growing, but the reality is that not every complex model can be fully explained in simple terms. Thus, the onus is on valuers to implement robust validation steps and maintain accountability for the final valuation outcome, even when leveraging opaque tools. Overall, the message is that high-quality data governance and strict quality assurance protocols are essential companions to new technology. By ensuring data are accurate and representative and by demanding clarity about how models operate, the profession can mitigate the risks of bias and error, thereby preserving public trust in valuations produced with the aid of AI and other advanced techniques.

Global context and equity considerations

While the above developments are largely drawn from well-resourced settings, the volume raises cautions about global disparities and the need for context-specific adaptation. Valuation technologies and standards developed in one country may not translate neatly to another. For instance, tools like advanced AVMs or drone surveys assume certain infrastructure, such as extensive sales databases, stable power and internet connectivity, and consistent legal frameworks, that may be lacking in developing markets. Wallace, Wallace, Logue and Moorhead (Chapter 6) note that even within a technologically advanced country, institutional readiness varies; the challenge is magnified in regions with informal property markets or limited regulatory oversight. Without mindful adaptation, there is a risk that technology could widen the gap between valuation practitioners in high-resource versus low-resource environments.

Moreover, the volume touches on issues of cultural fit and epistemic diversity. Professional standards and analytic methods shaped by Western practice might overlook traditional or local valuation approaches. In some communities, for example, Indigenous contexts or places where property rights are governed by custom and the notions of value and acceptable evidence differ from the assumptions built into mainstream valuation models. An algorithm trained on global financial data might not account for culturally significant attributes or communal forms of ownership. This highlights the importance of inclusive design and global dialogue: the profession must strive to ensure that technological advancement does not marginalise valid alternative perspectives on value. International valuation forums and bodies are beginning to discuss how to accommodate different epistemologies and contexts so that modernisation efforts benefit a broad spectrum of markets rather than imposing a one-size-fits-all model. Khanjanasthiti, Gardner and Taklif (Chapter 3) demonstrate through case studies of Landgate in Western Australia and Nova Scotia's PVSC (Property Valuation Services Corporation) how statutory authorities with commercial governance structures can successfully integrate AI and ML technologies while maintaining regulatory compliance.

The volume highlights that while AI and other advanced valuation tools have seen successful deployment in well-regulated, data-rich markets, their adoption in low-resource or developing contexts remains a substantial challenge. In such environments, even basic valuation processes may be constrained by limited infrastructure and data availability, indicating that digital tools cannot be introduced effectively without first addressing essential institutional and technical foundations (Skulmoski, Lang, Memari and Lee, Chapter 13). It also notes that different professional jurisdictions, like those under European or North American valuation standards, may diverge in their pace of change. Ensuring some harmony in how various standard-setters embrace technology will be important for maintaining a globally consistent level of professionalism and avoiding a fragmented landscape of practice.

Practical implementation pathways and barriers

Implementing valuation technology is not done by simply flipping a switch. The volume shows that organisations go through iterative phases when adopting new tools, learning and adjusting at each step. Broadly, the process can be distilled into three stages:

- **Assessment (Phase 1)**: This stage involves identifying which technologies align with organisational needs, operational capabilities and regulatory requirements. It includes reviewing available tools and practices, analysing cost and benefit considerations and consulting stakeholders such as valuers, clients and information technology (IT) staff. For example, a firm may consider using an AI-powered AVM for selected property types or explore drones for remote inspections. The outcome of this phase should be a clear implementation plan that outlines anticipated benefits, risks and any gaps in skills, infrastructure or policy that need to be addressed.
- **Pilot (Phase 2)**: The selected technology is deployed on a limited scale to test performance in practice. Chapters describe examples such as trialling drone inspections on a few sites alongside traditional methods or using AI tools in parallel with conventional valuation workflows. Clear criteria are used to evaluate performance, such as whether AI-generated figures align with professional judgement or whether drone use improves inspection speed without reducing accuracy. This stage also allows for the identification of issues,

including the risk of generative AI producing biased or unreliable outputs if prompts are poorly constructed. Feedback from users and clients supports refinement of systems and processes. The pilot concludes with a decision to adopt, modify or abandon the technology.

- **Integration (Phase 3)**: If the results are positive, the technology is integrated into routine operations. This includes full deployment, staff training and updates to procedures. Interoperability becomes a central focus to ensure new systems work smoothly with existing infrastructure. For example, data from a LiDAR scan or drone should feed directly into the firm's valuation reporting software. Organisations may assign internal technology leads to guide uptake, support colleagues and address issues. At this stage, the innovation becomes part of daily practice, supported by consistent application and internal oversight.
- **Continuous monitoring (Phase 4)**: Ongoing review ensures the technology is used appropriately and remains effective. Formal protocols may be introduced, such as requiring a valuer to verify AI-generated outputs before inclusion in a report or ensuring drone imagery is archived for compliance. As integration matures, attention turns to maintaining systems, updating tools and aligning with regulatory developments. When well managed, the technology becomes a stable component of professional valuation work.

Throughout these phases, the chapters highlight a range of challenges that can impede implementation if not addressed:

- **Regulatory and standards constraints**: Professional standards such as the International Valuation Standards often include requirements that limit the scope for full automation. For example, some standards specify that a valuation must include a physical inspection. Until these regulations evolve, for instance, by allowing a virtual inspection using drone imagery under specified conditions, valuers must deploy technology in ways that remain compliant with existing rules. This typically means using technology to supplement, rather than replace, established processes. Strict licensing frameworks or client-specific mandates may further delay the adoption of new methods.
- **Data and infrastructure limitations**: Advanced valuation technologies depend on the availability of reliable data and digital infrastructure. In some markets, the absence of comprehensive digitised property records or stable internet connectivity presents a major barrier. An AI model with insufficient data or a drone operating in areas without a GPS signal cannot deliver reliable outputs. Additionally, data privacy regulations may restrict the sharing of information required for AI model development and training.
- **Cost and training demands**: The initial costs associated with adopting new technologies can be high. These may include equipment purchases, software licensing and the allocation of time and resources for staff training (Neuger and Armitage, Chapter 10). Smaller firms may lack the capacity to invest in

technologies such as LiDAR or to recruit personnel with data science expertise. Learning new systems can temporarily reduce productivity, especially during the adjustment period. Khanjanasthiti, Taklif and Gardner (Chapter 7) detail how Australia's comprehensive drone regulations, while ensuring safety, can create 70-day approval processes that significantly delay technology implementation even when benefits are clearly demonstrated. Without adequate training and sustained support, even effective tools may remain underused or misapplied.

- **Cultural resistance and trust**: Organisational culture and individual attitudes significantly influence the adoption of innovation. Some valuers may express scepticism about algorithmic outputs or perceive technology as a threat to their professional judgement. Case examples highlight instances where experienced practitioners distrusted AI-generated recommendations or viewed remote inspections as inferior to on-site visits. Building trust requires clear communication about how new tools operate and how they complement, rather than undermine, professional expertise. Leadership support is critical to managing change and addressing the concerns of clients who are cautious of technology-enabled valuation components with underlying algorithms that are not transparent. Managing these expectations and explaining the role of technology in the valuation process is essential to gaining acceptance.
- **Liability and ethical concerns**: Uncertainty about liability presents another obstacle. Valuers may question who bears responsibility if a drone fails to detect a critical defect or an AI-generated valuation proves significantly inaccurate. Without established frameworks for accountability, whether through professional guidance or insurance mechanisms, practitioners may be reluctant to rely on automated tools. Ethical considerations also come into play, including the need to prevent algorithmic bias and ensure fairness. These concerns may lead organisations to adopt a cautious, incremental approach to integration.

Encouragingly, the case studies also show that these barriers can be overcome with proactive strategies. For instance, Cradduck (Chapter 11) detailed how a collaborative approach between regulators, industry and academia helped to introduce an AI tool within an existing legal framework, demonstrating that flexibility and open dialogue can smooth the adoption process. In contrast, Wallace, Wallace, Logue and Moorhead (Chapter 6) showed how a public-sector valuation unit's rigid procedures and risk aversion significantly delayed use of a proven drone inspection system, underlining that institutional inertia can stifle innovation even when its benefits are clear. In general, early engagement with policymakers, investment in thorough training, starting small with controlled pilots and fostering a culture of innovation all contribute to a successful transition. The key is acknowledging the obstacles upfront and addressing them through adaptive policies and continuous feedback, rather than abandoning the technological opportunity altogether.

Implications for professional development and quality assurance

As technology becomes ingrained in valuation practice, education and professional development are evolving to keep pace. New graduates entering the field are now expected to have exposure to data analytics, geospatial technology and even basic programming, alongside traditional valuation theory. Boyd (Chapter 9) emphasises that education reforms must balance technical competency development with critical thinking skills, ensuring graduates can not only operate new tools but also maintain professional judgement in an AI-augmented environment. For example, valuation courses are incorporating modules on automated valuation models and spatial data analysis, and professional bodies are introducing continuing professional development modules on topics like AI ethics or drone operation. Neuger and Armitage (Chapter 10) stress that continuing professional development programmes must evolve to address mid-career transitions, helping experienced valuers adapt their established expertise to incorporate new technological capabilities effectively. A recurring point is that training must emphasise not just how to use a tool but also how to interpret its outputs and remain compliant with professional ethics and standards while doing so.

Mid-career seminars and workshops are emerging that present case studies – for instance, how to handle a situation where an AVM's estimate conflicts with market evidence – to reinforce the valuer's role in exercising judgement and maintaining accountability in an AI-assisted workflow. Wolf and Amar (Chapter 8) propose an 'ethical AI' framework encompassing awareness, explainability, fairness, transparency and accountability as essential pillars for training valuers to use new tools responsibly while understanding their potential biases and limitations. At the same time, formally updating qualification criteria to include data science or coding competencies is still ongoing; professional institutions are gradually figuring out how to certify these skills as part of valuation expertise.

Quality assurance mechanisms and professional standards are likewise adapting to the digital era. Existing valuation standards such as the International Valuation Standards set by the IVSC (International Valuation Standards Council) or RICS's global Red Book are principle-based, meaning they do not ban the use of specific technologies but require that valuations, however conducted, are carried out with competence, transparency and due care (Crawford & Brogan, Chapter 2). In practice, this means that if a valuer uses an AI model or a drone, they must still meet all the usual reporting and evidence requirements. Many contributors stress the importance of documenting and explaining the use of technology in valuation reports. For example, if an AVM was consulted, the report should note how it was used and any steps taken to validate its output against the valuer's own analysis.

If a drone was used for inspection, any limitations, such as areas not visible by the drone, should be acknowledged and supplemented by other verification.

Some leading firms have developed internal guidelines and essential quality checklists to govern the use of AI and remote sensing tools and ensure that a qualified valuer reviews and signs off on all key outputs and that data from new tools is archived for audit and review. Professional bodies are also beginning to issue guidance on these matters: there are international discussions about what constitutes an 'adequate inspection' or an 'adequate explanation' when using AI, and there is regulatory interest, for example, the European Union's proposed AI Act, in mandating transparency and risk controls for high-stakes algorithms.

Notably, leading professional bodies have begun to address these issues collaboratively. RICS and other organisations have set up working groups to examine how standards might accommodate AI and remote inspection techniques, and international forums are facilitating the sharing of best practices. Such efforts aim to ensure a coordinated evolution of standards worldwide in response to technological change. Regulators are adopting an increasingly anticipatory stance by trying to address potential issues before they manifest, rather than reacting after the fact. As the profession embraces ever more complex tools, these core principles and safeguards serve as vital anchors to ensure valuations remain credible and defensible.

Future outlook and research directions

Throughout the case studies, it is evident that the tensions between new technology and established practice involving automation versus judgement or efficiency versus equity can become catalysts for improvement if managed proactively. Rather than eliminating these frictions, the profession can channel them into better standards, training and methods. Looking ahead, the valuation profession will need to continue adapting and actively shape its future. The following are some broad areas where further research and proactive efforts are needed to ensure technology is integrated in a way that strengthens, rather than undermines, the profession:

- **Ethical and equitable AI integration**: Developing frameworks to use AI and algorithms at scale without compromising fairness, transparency or public trust. This includes global cooperation among professional bodies and regulators to establish guidelines for algorithmic valuation that prevent bias and require appropriate human oversight. An important research question is how to build AI systems that valuers and stakeholders can trust and systems that improve efficiency while upholding ethical standards and serving all market segments fairly.
- **Preserving professional expertise and accountability**: Ensuring that as more valuation tasks become automated, the role of the valuer remains one of an authoritative decision-maker. Further study is needed on how to maintain professional agency in an AI-rich workflow; for example, by designing

decision-support tools that explain their reasoning or by delineating clear responsibility when humans and machines collaborate. The challenge is to avoid a deskilling of practitioners; instead, technology should augment their expertise. New models of professional accountability may be required so that liability and credit are appropriately assigned when AI is involved in valuations.

- **Embracing emerging technologies and data complexities**: Exploring the implications of frontier technologies like quantum computing, advanced sensors or real-time big data feeds for valuation. These innovations could enable analysis of asset markets at an unprecedented scale and speed, but they will also introduce new complexities. Skulmoski, Lang, Memari and Lee (Chapter 13) explore how frontier technologies like quantum computing could enable unprecedented scale and speed in asset market analysis while introducing new complexities requiring early establishment of validation standards and professional guidelines. Anticipating these developments will help the profession set standards and best practices early rather than reacting after the fact.

By actively participating in the design and governance of new systems, valuers can help ensure that innovation is guided by professional values and the public interest, rather than treating technology as an external inevitability. The case studies in this volume demonstrate that with thoughtful implementation and a commitment to continual learning, technology can be a powerful ally. Although these discussions focus on real estate valuers, the underlying lesson of balancing innovation with professional responsibility resonates across other knowledge-based fields, from medicine to law to engineering. Valuers can thereby provide deeper insights and more efficient services without eroding the ethical and professional foundations on which public confidence has long been built.

Index

Pages in *italics* refer to figures and pages in **bold** refer to tables.

airports 127, 129, 131, 133, **135**, 137, 139, 141–142, 145–146
Apple Vision Pro 219–222, 227, 231
apply, evaluate and solve (AES) model, educational 185–186, 192, 212
artificial intelligence (AI) 1, 12, 58, 60, 72, 74, 76, **78**, 102, 114, 125, 130, 132, 145, 147, 149, *168*, 170, 176, 178, 181, 186, 188, 190, 192, 194, 196, *198*, 216, 227–228, 238, 245, 270, 272, 274, 277, 279, 283, 285, 291
asset management 109, 111–114, 124, 131, 149, 227
augmented professionalism 2, 7–9, 283
augmented reality (AR) 112, 219, 221, 223–224, 231
Australian Property Institute (API) 17–19, 30, 41, 54, 98, 199, 213, 218, 228, 252
automated valuation models (AVMs) 10, 12–13, 30, 35–37, 53–56, 60, 73, 77–81, 89, 97, 99, 102, 105, 113–114, 126, 218, 242, 253, 256–257, 265, 277, 280–281, 289; accuracy issues **78**, 79, 99; limitations 77, **78**, 253; regulatory concerns 113–114, 281

big data 4, 8, 12, 55–56, 72, 75, 101, 103, 215, **228**, 247–249, 251, 253–255, 257, 259, 265–266, *267*, 268, *269*, 291
blockchain 24, 100, **228**, 229, 261, 270, 274
building information modeling (BIM) 111

ChatGPT 77, **79**, 126, 239–242, 244–245, 262
childcare centres 74
climate change 21, 40
commercial property 72–73, 80, 87–89, 91, 98, 100, 219, 250
continuing professional development (CPD) 43, 123, 289

data analytics 64, 75, 150, 253, 282, 289
data privacy 42–43, 46, 66, 97, 121, 201, 271, 287
deep learning (DL) 33, 75, 91, 131–132, 143, **145**, 146, *151*, 187, 201
digital technologies, emerging (in valuation) 33, 44, 49, 127, 131, 146, 201, 253
digital twin(s) 111–112, 230, 249, 251, 282
discounted cash flow (DCF) 4–5, 10, 125, 245–246, 250
drones (unmanned aerial vehicles) 2–3, 6, 8, 110–111, 113, 122, 125–149, 186, 194–196, *197*, 203, **204–205**, 212, 214–215, 217, 263, 280–283, 285–289

education, valuation (education/training) 4, 7–9, 12, 66–70, 80, 183–233, 254, 256
environmental, social and governance (ESG) **74**, 100, 114
ethics 21, 29, 32, 45–46, 54, 68, 77, 83, 95, 97–100, 122, 207, **208**, 215, 255, 284, 289; AI and valuation 21, 29, 32, 45–46, 54, 68, 77, 83, 95, 97–100, 122, 207, **208**, 255, 289; ethical considerations 45, 99, 107, 118, 122, 140, 255, 288; professional ethics 21, 29, 54, 68, 83, 95, 98, 100, 122, 215, 255, 289

generative AI (Gen AI) 8, 24–26, 29–33, 45–46, 54–55, **79**, 150–159, 161–167, 169, 171–175, 177, 179, 181–182, 185–186, 239, 242–244, 246
geographic information systems (GIS) 113, 143
geolocation data (GPS) 111, 113, 287
governance 1, 9–10, 21, 45–48, 51, 54, 65–69, **74**, 76, 99, 107, 114, 118, 125, 206–210, 215, 255, 278, 281, 285–286, 291; algorithmic governance 1, 9, 46, 207, 215, 285

healthcare property (hospitals) 73–74, 223, 262, 265
hospitality property (hotels) 74, 259

industrial strata units 89, 91
International Valuation Standards (IVS) 4, 18, 22, 41, 107, 218, 226, 228, 231, 287, 289
International Valuation Standards Council (IVSC) 4, 9, 11–12, 19, 30, 35, 41, 55, 87, 107, 238, 289
Internet of Things (IoT) 4, 61, 152, **229**

Julius AI 151, 156–169, 171–175, 181

liability (professional) 8, 152, 255
LiDAR (light detection and ranging) 2–3, 6, 111, 122, 126, 131, 186, 194–197, *198*, 203, 212, 217, 222, 263, 279–283, 287–288

machine learning (ML) 1–2, 5–7, 18, 22, 27, 30, 33–53, 55–56, 62–63, 71–73, 75, 81, 83, 87, 98, 100–102, 111–112, 115–117, 121, 123, 151, 215, 248, 254, 268, 282
market analysis 219, 226, 265, 291
market forecasting 73, 258, **273**
market valuation 36, 52, 65, 76, 93, 115, 259
market value 18, 35–36, 38, 65, 87, 95, 125, 177, 180, 218, 223, 230, 248
mass appraisal (models) 3, 26, 35, 42, 48–50, 52, 81, 238, 281–282

NABERS **74**; *see also* sustainability

planning controls 96, 248
property market(s) 4, 22, 38, 67, 73, 93, 95, 153, 218, 262, 280, 285
property valuation 1–4, 9–10, 23–24, 34–37, 39–47, 49–50, 52, 57, 59, 67, 72–77, 80–84, 97, 99–100, 170, 218, 220, 223–224, 230, 247–248, 250–252, 255, 259, *267*; *see also* valuation approaches; valuation methods
PropTech (property technology) 11, 30–31, 83, 201, 214, 216, 220, 227, **228**, 229, 233, 247–248, 250

Q-Day (quantum computing milestone) 261–262, 273, 276
quantum computing 8, 229, 258–259, 262–266, 268, 270, 272–275
quantum key distribution (QKD) 263–264, 276–277

Red Book (RICS Valuation Standards) 86, 258, 289
remote sensing 281, 283–284, 290; *see also* drones; LiDAR; satellite imagery
residential property 54, 73, 83, 156, 216
retail property 248
risk management 3, 5, 8–9, 20, 40, 73, 80, 110, 122, 249, 255, 261, 271, 273
Royal Institution of Chartered Surveyors (RICS) 4–5, 36, 42, 86, 98, 253–254, 258, 290
runway inspections (airport) 8, 127–133, 137–141, 143, 145–146

satellite imagery 114, 248
spatial computing 229, 233; *see also* augmented reality (AR); virtual reality (VR)
statutory valuations 26, 51, 115, 117
sustainability 33–34, 39, 42, 48, 84, 100, 109, 118, 121, **207**, 210, 226, 254, 258–259, 272; *see also* environmental, social and governance (ESG)

valuation approaches 155, 192, 285; cost approach 177, 196; income approach 177; market approach 163, 177
valuation methods, traditional vs automated 4, 12–13, 43, 53–55, 101, 190, 218, 247
virtual reality (VR) 112, 164, 195, 221, 223–224, 228–230

XR (extended reality) 221, 223–224, 228

zoning 3, **74**, 84, 114, 219, 225, 227, 248, 251

For Product Safety Concerns and Information please contact our EU representative GPSR@taylorandfrancis.com Taylor & Francis Verlag GmbH, Kaufingerstraße 24, 80331 München, Germany